Periodic Character and Patterns of Recursive Sequences

Michael A. Radin

Periodic Character and Patterns of Recursive Sequences

 Springer

Michael A. Radin
School of Mathematical Sciences
Rochester Institute of Technology
Rochester, NY, USA

ISBN 978-3-030-01779-8   ISBN 978-3-030-01780-4   (eBook)
https://doi.org/10.1007/978-3-030-01780-4

Library of Congress Control Number: 2018960765

Mathematics Subject Classification: 11J71, 00A06, 03B70, 03C13, 03D20, 03F52, 03F50, 03F55, 03F60, 11B37, 11Y55, 34Kxx, 34K11, 34K13, 35F05, 35F20, 35F35, 35F50, 39–XX, 39Axx

This Springer imprint is published by the registered company Springer Nature Switzerland AG
The registered company address is: Gewerbestrasse 11, 6330 Cham, Switzerland

Preface

Understanding structures and details of patterns and repetitions is an essential part of the learning process that enhances our intuitions, analytical skills, and our deducting reasoning. For instance, it is essential to decipher the patterns and their repetitions when studying the weather systems, when learning to play a musical instrument, when studying a foreign language, when learning computer programming, when studying transportation routes and schedules, when analyzing engineering structures, and when studying behaviors and other similar applications.

Investigation of patterns is often claimed in medical sciences to detect particular viruses and spread of diseases, studying the heartbeats and frequencies of occurrences. Moreover, patterns are very frequently applied in signal processing and other analogous phenomena with waves and periodic traits.

The aim of this interdisciplinary book is to understand and recognize formations of periodic patterns analytically and graphically. We will gradually gain experience and knowledge of periodic traits by addressing the following questions:

- What is the period of the given cycle?

- Even ordered period or an odd ordered period?

- Existence and uniqueness of periodic cycles? Are periodic cycles unique or is every solution periodic?

- What is the relationship between the neighboring terms of the cycle?

- Are all the terms of the cycle positive or do terms alternate?

- Parity of neighboring terms?

- What are the necessary and sufficient conditions for the existence and uniqueness of periodic cycles?

- When eventually periodic solutions arise and why?

I invite you to the discovery journey in deciphering discrete patterns of first-order, second-order, and third- and higher-order difference equations. We will study the periodic traits of linear difference equations, rational difference equations, piece-wise difference equations, and max-type difference equations and discuss various applications. Our plan is to develop inductive intuition that will help us recognize specific structures of patterns of periodic cycles and eventually periodic cycles and develop theorems after numerous repetitive types of examples. The intents are to widen our inductive reasoning skills and develop techniques of proof by induction, proof by contradiction, and how additional keynotes such as number theory, com-binatorics, and abstract algebra and analyzing computer observations will blend in as pieces of the puzzle to address deeper research questions and to welcome the in-terdisciplinary research atmosphere. The aim is to enhance efficiency in fast-speed computing when we write computer programs that decipher the patterns of the pe-riodic cycles and the transient terms inductively that will lead to new unanswered questions to pioneer. This textbook can be applied in the following courses: In-troduction to Difference Equations, Advanced Discrete Mathematics, Vibrations, Resonance, Digital Systems, and Decision-Making.

Rochester Institute of Technology   Michael A. Radin
Rochester, New York

Contents

About the Author

Michael A. Radin earned his Ph.D. at the University of Rhode Island in 2001 and is currently an associate professor of mathematics at the Rochester Institute of Technology. He started his journey analyzing difference equations with periodic solutions as part of his Ph.D. thesis and has several publications on boundedness and periodic nature of solutions of rational difference equations, max-type difference equations, and piece-wise difference equations. Michael published several papers together with his Master's students and undergraduate students at RIT and has publications with students and colleagues from Riga Technical University and University of Latvia.

Michael has publications in applied mathematics-related topics such as Neural Networking, Modeling Extinct Civilizations, and Modeling Human Emotions. Also, Michael organized numerous sessions on difference equations and applications at the annual *American Mathematical Society* meetings. Michael participates actively in the annual *International Society of Difference Equations and Applications* and *Mathematical Modelling and Analysis* conferences. Recently, Michael published two manuscripts on international pedagogy and has been invited as a keynote speaker at several interdisciplinary conferences. Michael taught courses and conducted seminars on these related topics during his spring 2009 sabbatical at the Aegean University in Greece and during his spring 2016 sabbatical at Riga Technical University in Latvia.

During his spare time, Michael spends time outdoors and is an avid landscape photographer. In addition, Michael is an active poet and has several published poems in the *Le Mot Juste*. Spending time outdoors and active landscape photography widens and expands Michael's understandings of nature's patterns and cadences.

Chapter 1
Introduction

1.1 Introduction

We observe patterns every day throughout our lives. For instance, traffic patterns as we are driving, musical patterns as we are playing an instrument, nature's patterns such as clouds' cadence and rhythm of waves as we are spending time outdoors, weather patterns as we are traveling, patterns as we are writing computer programs, when studying psychological behavioral patterns, and mathematical patterns as we are solving particular math problems. Sometimes patterns repeat at the same scale and at other times patterns repeat at different scales. Here are the following examples that I would like to share of repeated patterns at the same scale and at different scales (Figures 1.1–1.4):

Fig. 1.1 Aerial photograph of the system of clouds with repeated patterns at the same scale taken by Michael A. Radin.

© Springer Nature Switzerland AG 2018
M. A. Radin, *Periodic Character and Patterns of Recursive Sequences*,
https://doi.org/10.1007/978-3-030-01780-4_1

Fig. 1.2 Photographs of the waves' patterns in the Gulf of Riga repeated at the same scale taken by Michael A. Radin.

Fig. 1.3 Aerial photograph of the Island of Patmos in Greece with repeated patterns at different scales taken by Michael A. Radin.

Our goals of this chapter and this textbook are to familiarize ourselves with difference equations (recursive sequences), to understand how to solve them explicitly using inductive reasoning, and to determine the patterns and properties of solutions and periodic cycles. To inductively understand, assemble and develop patterns after many repetitive types of examples which will then lead to discovery of theorems. To recognize whether patterns alternate or not, the relationship between the indices

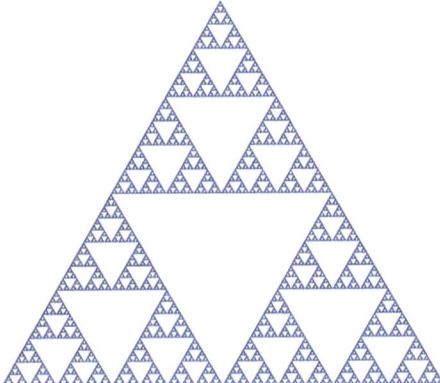

Fig. 1.4 Sierpinski Triangle depicting a system of equilateral triangles at different scales. This image was produced by James W. Wilson from the University of Georgia.

from neighbor to neighbor and to compare the similarities and difference between even ordered periodic cycles and odd ordered periodic cycles and other deeper traits. Discover the contrasts when we have unique periodic cycles versus every solution is periodic. Besides, we will establish the patterns of the periodic cycles with respect to even ordered indices and odd ordered indices of a periodic sequence and how they change or rotate from neighbor to neighbor. We will illustrate examples of such patterns graphically and analytically at the end of the chapter and contrive such patterns throughout the book. Study particular difference equations that exhibit eventually periodic solutions and compare the similarities and difference between periodic solutions and eventually periodic solutions. Why do some particular difference equations or recursive sequences have transient behavior?

1.2 Recursive Sequences

We will start refreshing our memories about sequences. First of all, how do we recognize the specific pattern and then write a formula of a sequence? For example, a sequence that lists multiples of 3, a geometric sequence, and a sequence that resembles the factorial pattern. Second, for what patterns must we treat as recursive and therefore write a recursive formula? When we write a recursive formula for a sequence, we will treat it as an **Initial Value Problem**, where the initial value is the starting value or the starting point of the sequence. We will start with examples when we move from one neighbor to the next by adding a specific constant. In the very first example we will move from one neighbor to the next by adding a 1; this will list all the positive integers starting at 1.

Example 1. Write a formula for:

$$1 , 2 , 3 , 4 , 5 , 6 , 7 , \dots .$$

Solution: Observe that we move from neighbor to neighbor by adding a 1 and we can write the following formula:

$$\{n\}_{n=1}^{\infty}$$

Alternatively, we can treat this as a recursive sequence. By iteration and induction we get:

$$x_1 = 1 ,$$

$$x_1 + 1 = 1 + 1 = 2 = x_2 ,$$

$$x_2 + 1 = 2 + 1 = 3 = x_3 ,$$

$$x_3 + 1 = 3 + 1 = 4 = x_4 ,$$

$$x_4 + 1 = 4 + 1 = 5 = x_5 ,$$

$$x_5 + 1 = 5 + 1 = 6 = x_6 ,$$

$$\vdots$$

Hence for all $n \in \mathbb{N}$:

$$\begin{cases} x_{n+1} = x_n + 1, \\ \\ x_1 = 1 . \end{cases}$$

Example 2. Write a formula for:

$$5 , 9 , 13 , 17 , 21 , 25 , 29 , \dots .$$

Solution: Observe that we move from neighbor to neighbor by adding a 4 and express the following formula:

$$\{4n + 1\}_{n=1}^{\infty}$$

Now by iteration and induction we get:

$$x_1 = 5 ,$$

$$x_1 + 4 = 5 + 4 = 9 = x_2 ,$$

$$x_2 + 4 = 9 + 4 = 13 = x_3 ,$$

$$x_3 + 4 = 13 + 4 = 17 = x_4 ,$$

$$x_4 + 4 = 17 + 4 = 21 = x_5 ,$$

$$x_5 + 4 = 21 + 4 = 25 = x_6 \,,$$

$$x_6 + 4 = 25 + 4 = 29 = x_7 \,,$$

$$\vdots$$

Thus for all $n \in \mathbb{N}$:

$$\begin{cases} x_{n+1} = x_n + 4 \,, \\ \\ x_1 = 5 \,. \end{cases}$$

In the next example we move from neighbor to neighbor by multiplying by 2. This is called a **Geometric Sequence**.

Example 3. Write a formula for:

$$3 \,,\, 6 \,,\, 12 \,,\, 24 \,,\, 48 \,,\, 96 \,,\, 192 \,,\ldots \,.$$

Solution: First of all, we obtain the following formula:

$$\{3 \cdot 2^n\}_{n=0}^{\infty} \,.$$

Second, we get:

$$x_1 = 3 \,,$$

$$x_1 \cdot 2 = 3 \cdot 2 = 6 = x_2 \,,$$

$$x_2 \cdot 2 = 6 \cdot 2 = 12 = x_3 \,,$$

$$x_3 \cdot 2 = 12 \cdot 2 = 24 = x_4 \,,$$

$$x_4 \cdot 2 = 24 \cdot 2 = 48 = x_5 \,,$$

$$x_5 \cdot 2 = 48 \cdot 2 = 96 = x_6 \,,$$

$$x_6 \cdot 2 = 96 \cdot 2 = 192 = x_7 \,,$$

$$\vdots$$

Hence for all $n \in \mathbb{N}$:

$$\begin{cases} x_{n+1} = 2x_n \,, \\ \\ x_1 = 3 \,. \end{cases}$$

In the previous three examples we wrote the **Initial Value Problem** as a solution. Furthermore, in the previous three examples we move from neighbor to neighbor by either adding or multiplying different terms.

Example 4. Write a recursive formula (as an Initial Value Problem) for:

$$1 \,,\, 2 \,,\, 4 \,,\, 7 \,,\, 11 \,,\, 16 \,,\, 22 \,,\ldots .$$

Solution: Notice:

$$x_1 = 1 \,,$$

$$x_1 + 1 = 1 + 1 = 2 = x_2 \,,$$

$$x_2 + 2 = 2 + 2 = 4 = x_3 \,,$$

$$x_3 + 3 = 4 + 3 = 7 = x_4 \,,$$

$$x_4 + 4 = 7 + 4 = 11 = x_5 \,,$$

$$x_5 + 5 = 11 + 5 = 16 = x_6 \,,$$

$$x_6 + 5 = 16 + 5 = 21 = x_7 \,,$$

$$\vdots$$

It follows by induction that for all $n \in \mathbb{N}$:

$$\begin{cases} x_{n+1} = x_n + n \,, \\ \quad x_1 = 1 \,. \end{cases}$$

Example 5. Write a recursive formula (as an Initial Value Problem) for:

$$4 \,,\, 19 \,,\, 39 \,,\, 64 \,,\, 94 \,,\, 129 \,,\ldots .$$

Solution: By iteration and induction we get:

$$x_1 = 4 \,,$$

$$x_1 + 5 \cdot 3 = 4 + 15 = 4 + 5 \cdot 3 = 19 = x_2 \,,$$

$$x_2 + 5 \cdot 4 = 19 + 20 = 19 + 5 \cdot 4 = 39 = x_3 \,,$$

$$x_3 + 5 \cdot 5 = 39 + 25 = 39 + 5 \cdot 5 = 64 = x_4 \,,$$

$$x_4 + 5 \cdot 6 = 64 + 30 = 64 + 5 \cdot 6 = 94 = x_5 \,,$$

$$x_5 + 5 \cdot 7 = 94 + 35 = 94 + 5 \cdot 7 = 129 = x_6 \,,$$

$$\vdots$$

We see that for all $n \in \mathbb{N}$:

$$\begin{cases} x_{n+1} = x_n + 5(n+2) , \\ x_1 = 4 . \end{cases}$$

Example 6. Write a recursive formula (as an Initial Value Problem) for:

$$6 , 8 , 12 , 20 , 36 , 68 , 132 , \dots .$$

Solution: By iteration and induction we get:

$$x_1 = 6 ,$$

$$x_1 + 2^1 = 6 + 2 = 6 + 2^1 = 8 = x_2 ,$$

$$x_2 + 2^2 = 8 + 4 = 8 + 2^2 = 12 = x_3 ,$$

$$x_3 + 2^3 = 12 + 8 = 12 + 2^3 = 20 = x_4 ,$$

$$x_4 + 2^4 = 20 + 16 = 20 + 2^4 = 36 = x_5 ,$$

$$x_5 + 2^5 = 36 + 32 = 36 + 2^5 = 68 = x_6 ,$$

$$\vdots$$

Thus for all $n \in \mathbb{N}$:

$$\begin{cases} x_{n+1} = x_n + 2^n , \\ x_1 = 6 . \end{cases}$$

Example 7. Write a recursive formula (as an Initial Value Problem) for:

$$1 , 2 , 6 , 24 , 120 , 720 , 5040 , \dots .$$

Solution: By iteration and induction we get:

$$x_1 = 1 ,$$

$$x_1 \cdot 2 = 1 \cdot 2 = 2 = x_2 ,$$

$$x_2 \cdot 3 = 2 \cdot 3 = 6 = x_3 ,$$

$$x_3 \cdot 4 = 6 \cdot 4 = 24 = x_4 ,$$

$$x_4 \cdot 5 = 24 \cdot 5 = 120 = x_5 ,$$

$$x_5 \cdot 6 = 120 \cdot 6 = 720 = x_6 ,$$

$$\vdots$$

Therefore, for all $n \in \mathbb{N}$:

$$\begin{cases} x_{n+1} = (n+1) \cdot x_n \,, \\ \\ x_1 = 1 \,. \end{cases}$$

Notice that this example resembles the pattern of a **factorial**.

1.3 Order of a Difference Equation and Explicit Solution

Now we will analyze examples of difference equations of various orders:

(i) $x_{n+1} = \frac{x_n}{x_n-1}$, $n = 0, 1, \ldots$.

(ii) $x_{n+2} = \frac{x_n}{1+x_{n+1}}$, $n = 0, 1, \ldots$.

(iii) $x_{n+3} = x_{n+2} + x_{n+1} + x_n$, $n = 0, 1, \ldots$.

In (i) we have a difference equation of first order as x_{n+1} depends on x_n. In (ii) we have a difference equation of second order as x_{n+2} depends on x_{n+1} and x_n. In (iii), we have a difference equation of third order as x_{n+3} depends on x_{n+2}, x_{n+1}, and x_n. Now we will define a first order difference equation (Δ.E. as an abbreviation) as an iterative process (recursive relation) in the form:

$$x_{n+1} = f(x_n) \quad , \quad n = 0, 1, \ldots \, . \tag{1.1}$$

Equation (1.1) is a **first order** Δ.E. with one initial condition x_0. The function $y = f(x)$ describes Equation (1.1) on an interval (domain) I. If $f : I \to I$ and $x_0 \in I$, then $x_n \in I$ for all $n \geq 0$. The following example portrays difference equations that are of first order as x_{n+1} depends on x_n.

Example 8. The following two difference equations are of first order:

(i)   (Special Case of the **Logistic Δ.E.**)

$$x_{n+1} = 4x_n(1-x_n) \quad , \quad n = 0, 1, \ldots \, .$$

(ii)   (Special Case of the **Riccati Δ.E.**)

$$x_{n+1} = \frac{x_n}{x_n - 1} \quad , \quad n = 0, 1, \ldots \, .$$

Now we will discuss a solution to Equation (1.1). Observe that by iterating Equation (1.1), we procure:

$$x_0, \ x_1 = f(x_0), \ x_2 = f(x_1) = f^2(x_0), \ \ldots \ , \ x_n = f^n(x_0), \ldots \, .$$

The solution to Equation (1.1) is defined as a **sequence** $\{x_n\}_{n=0}^{\infty}$. Our aims are to explicitly solve difference equations by iterations (recursively) and inductively as we will see throughout the book.

Example 9. Determine the **explicit solution** of:

$$x_{n+1} = 3x_n \quad , \quad n = 0, 1, \dots .$$

Solution: By iteration and induction we get:

$$x_0 \, ,$$

$$x_1 = 3x_0,$$

$$x_2 = 3x_1 = 3 \cdot [3x_0] = 3^2 x_0 \, ,$$

$$x_3 = 3x_2 = 3 \cdot [3^2 x_0] = 3^3 x_0 \, ,$$

$$x_4 = 3x_3 = 3 \cdot [3^3 x_0] = 3^4 x_0 \, ,$$

$$x_5 = 3x_4 = 3 \cdot [3^4 x_0] = 3^5 x_0 \, ,$$

$$\vdots$$

We see that for all $n \in \mathbb{N}$:

$$x_n = 3^n x_0 \, .$$

Example 10. Determine the **explicit solution** of:

$$x_{n+1} = 2x_n^2 \quad , \quad n = 0, 1, \dots .$$

Solution: By iteration and induction we get:

$$x_0 \, ,$$

$$x_1 = 2x_0^2 \, ,$$

$$x_2 = 2x_1^2 = 2 \left[2x_0^2 \right]^2 = 2^3 x_0^4 \, ,$$

$$x_3 = 2x_2^2 = 2 \left[2^3 x_0^4 \right]^2 = 2^7 x_0^8 \, ,$$

$$x_4 = 2x_3^2 = 2 \left[2^7 x_0^8 \right]^2 = 2^{15} x_0^{16} \, ,$$

$$x_5 = 2x_4^2 = 2 \left[2^{15} x_0^{16} \right]^2 = 2^{31} x_0^{32} \, ,$$

$$\vdots$$

Hence we see that for all $n \in \mathbb{N}$:

$$x_n = 2^{(2^n-1)} x_0^{2^n} .$$

Next we will examine the Δ.E. in the form:

$$x_{n+2} = f(x_{n+1}, x_n) \quad , \quad n = 0, 1, \dots . \tag{1.2}$$

Notice that this is a **second order** Δ.**E.** as we have two initial conditions x_0 and x_1 and x_{n+2} depends on x_{n+1} and x_n. In addition, the function $f(x, y)$ describes Equation (1.2) on an interval (domain) $I \times I$. If $f : I \times I \to I \times I$ and $x_0, x_1 \in I$, then $x_n \in I$ for all $n \geq 0$.

Similar to Equation (1.1), we see that a solution to Equation (1.2) is a sequence $\{x_n\}_{n=0}^{\infty}$.

From Equations (1.1) and (1.2) we extend to the kth ($k \in \mathbb{N}$) order Δ.E.:

$$x_{n+k} = f(x_{n+k-1}, x_{n+k-2}, \dots, x_n) \quad , \quad n = 0, 1, \dots , \tag{1.3}$$

with k initial conditions $x_0, x_1, \dots, x_{k-1}$, whose solution is a sequence $\{x_n\}_{n=0}^{\infty}$.

Our aim is to investigate the **long-term behavior** of solutions to difference equations. It is of paramount interest to address the following questions:

- Does every solution have a limit?

- Do periodic solutions exist?

- Is every solution bounded?

The primary goal of this textbook is to study the necessary and sufficient conditions for the existence and uniqueness of periodic solutions, their patterns, and other vital properties. Now we will introduce the **Equilibrium Point** as a potential property of difference equations.

1.4 Equilibrium Points

We say that $\bar{x}$ is an **Equilibrium Point** of Equation (1.1) provided that

$$\bar{x} = f(\bar{x}) .$$

If $x_0 = \bar{x}$, then $x_n = \bar{x}$ for all $n \geq 0$. In addition, we say that $\bar{x}$ is a **trivial solution** of Equation (1.1). Now we analyze the following examples.

Example 11. Determine the **equilibrium point(s)** of:

$$x_{n+1} = \frac{4x_{n-1}}{1 + x_n} \quad , \quad n = 0, 1, \dots .$$

Solution: Note, we set

$$\bar{x} = \frac{4\bar{x}}{1+\bar{x}},$$

from which we see that $\bar{x} = 0$ and $\bar{x} = 3$ are the only equilibrium points of the given Δ.E.

Example 12. Determine the **equilibrium point(s)** of:

$$x_{n+2} = x_{n+1}e^{-x_n} \quad , \quad n = 0, 1, \ldots .$$

Solution: First we set

$$\bar{x} = \bar{x}e^{-\bar{x}},$$

from which we see that $\bar{x} = 0$ is the only equilibrium point of the given Δ.E.

From Example 11 and Example 12, it is vital to determine if $\lim_{n\to\infty} x_n = \bar{x}$? We will examine examples that address this question in Chapter 2.

Now we will shift our focus on **periodic sequences** and on difference equations with periodic solutions and their patterns.

1.5 Periodic Sequences (Solutions)

Now our aim is to recognize periodic traits and patterns graphically and analytically. In Pre-Calculus, we studied periodic functions such as $y = \sin(x)$, $y = \cos(x)$, and Piece-wise Functions as $y = |x|$. The graphs below are examples of a cosine function with period-50 and of a Piece-wise Function with period-4 (Figures 1.5 and 1.6):

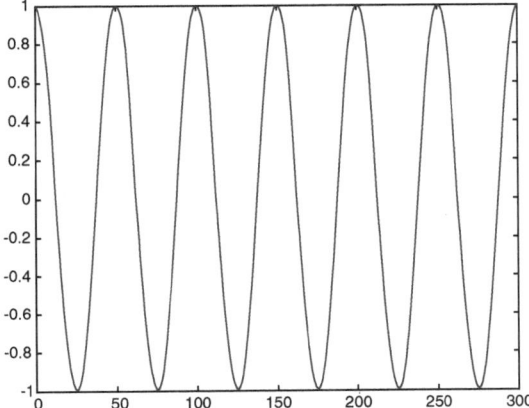

Fig. 1.5 Graph of a cosine function with period-50 and depicts 6 cycles.

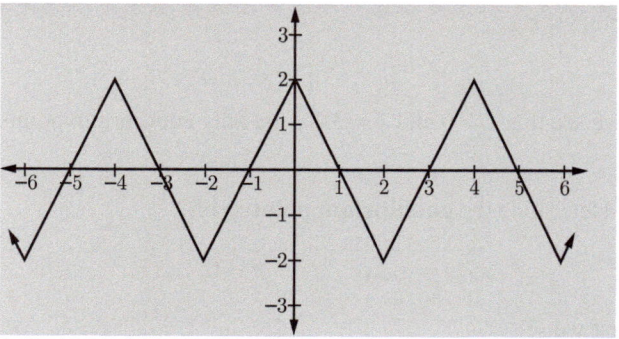

Fig. 1.6 Graph of a Piece-wise function with period-4 and depicts 3 cycles.

The sequence $\{x_n\}_{n=0}^{\infty}$ is periodic with **minimal period-p,** $(p \geq 2)$, provided that

$$x_{n+p} = x_n \quad \text{for all} \quad n = 0, 1, \ldots .$$

The next two graphs are examples of a period-3 cycle and a period-6 cycle (Figures 1.7 and 1.8):

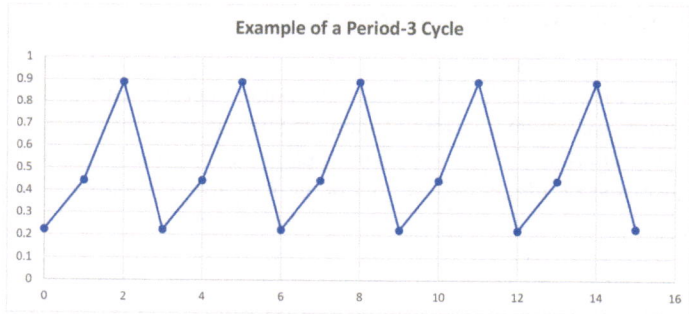

Fig. 1.7 Graph of a period-3 cycle with increasing terms of the **Tent-Map**.

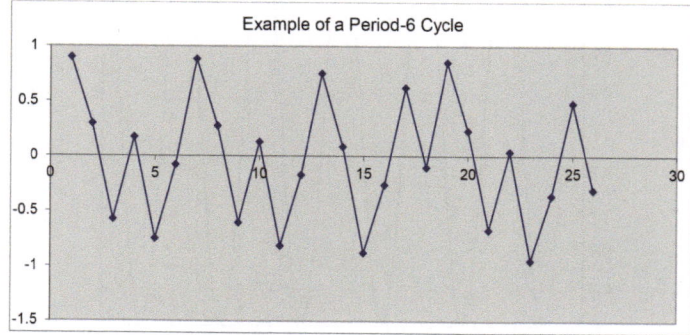

Fig. 1.8 Graph of an oscillatory period-6 cycle of a **piece-wise difference equation**.

In Examples 1–7 we wrote an **Initial Value Problem** as a solution. Now the upcoming examples will determine the patterns of different periodic cycles analytically by solving the given **Initial Value Problem**.

Example 13. Solve the **Initial Value Problem** explicitly and determine the period.

$$\begin{cases} x_{n+1} = -x_n + 8 \, , \quad n = 0, 1, \dots , \\ \\ x_0 = 2 \, . \end{cases}$$

Solution: Notice:

$$x_0 = 2 \, ,$$

$$x_1 = -x_0 + 8 = -2 + 8 = 6 \, ,$$

$$x_2 = -x_1 + 8 = -6 + 8 = 2 = x_0 \, ,$$

$$x_3 = -x_2 + 8 = -2 + 8 = 6 = x_1 \, .$$

We obtain a period-2 cycle and for all $n \geq 0$:

$$\begin{cases} x_{2n} = x_0 = 2 \, , \\ \\ x_{2n+1} = x_1 = 6 \, . \end{cases}$$

Example 14. Solve the **Initial Value Problem** explicitly and determine the period.

$$\begin{cases} x_{n+1} = \frac{1}{x_n} \, , \quad n = 0, 1, \dots , \\ \\ x_0 = 4 \, . \end{cases}$$

Solution: By induction we get:

$$x_0 = 4 \, ,$$

$$x_1 = \frac{1}{x_0} = \frac{1}{4} \, ,$$

$$x_2 = \frac{1}{x_1} = \frac{1}{\frac{1}{4}} = 4 = x_0 \, ,$$

$$x_3 = \frac{1}{x_2} = \frac{1}{4} = x_1 \, .$$

Hence we acquire a period-2 cycle and for all $n \geq 0$:

$$\begin{cases} x_{2n} = x_0 = 4 \, , \\ \\ x_{2n+1} = x_1 = \frac{1}{4} \, . \end{cases}$$

Example 15. Solve the **Initial Value Problem** explicitly and determine the period.

$$\begin{cases} x_{n+2} = -x_n \, , \quad n = 0, 1, \dots \, , \\ \\ x_0 = 1 \, , \\ \\ x_1 = 5 \, . \end{cases}$$

Solution: Observe:

$$x_0 = 1 \, ,$$

$$x_1 = 5 \, ,$$

$$x_2 = -x_0 = -1 \, ,$$

$$x_3 = -x_1 = -5 \, ,$$

$$x_4 = -x_2 = 1 = x_0 \, ,$$

$$x_5 = -x_3 = 5 = x_1 \, .$$

Hence we see a period-4 cycle and for all $n \geq 0$:

$$\begin{cases} x_{4n} = x_0 = 1 \, , \\ \\ x_{4n+1} = x_1 = 5 \, , \\ \\ x_{4n+2} = x_2 = -1 \, , \\ \\ x_{4n+3} = x_3 = -5 \, . \end{cases}$$

Observe that this is an alternating period-4 pattern.

Example 16. Solve the **Initial Value Problem** explicitly and determine the period.

$$\begin{cases} x_{n+2} = -x_{n+1} - x_n \, , \quad n = 0, 1, \dots \, , \\ \\ x_0 = 7 \, , \\ \\ x_1 = 3 \, . \end{cases}$$

Solution: By iteration:

$$x_0 = 3,$$

$$x_1 = 7,$$

$$x_2 = -x_1 - x_0 = -7 - 3 = -10,$$

$$x_3 = -x_2 - x_1 = 10 - 7 = 3 = x_0,$$

$$x_4 = -x_3 - x_2 = -3 + 10 = 7 = x_1.$$

Hence we see a period-3 cycle and for all $n \geq 0$:

$$\begin{cases} x_{3n} = x_0 = 3, \\ x_{3n+1} = x_1 = 7, \\ x_{3n+2} = x_2 = -10. \end{cases}$$

Example 17. Solve the **Initial Value Problem** explicitly and determine the period.

$$\begin{cases} x_{n+1} = a_n x_n, \quad n = 0, 1, \ldots, \\ x_0 = 1, \end{cases}$$

where $\{a_n\}_{n=0}^{\infty}$ is a period-2 sequence defined as:

$$a_n = \begin{cases} a_0 \text{ if n is even,} \\ a_1 \text{ if n is odd.} \end{cases}$$

and assume $a_0 a_1 = 1$.

Solution: Notice:

$$x_0 = 1,$$

$$x_1 = a_0 x_0 = a_0,$$

$$x_2 = a_1 x_1 = a_1 a_0 = 1 = x_0,$$

$$x_3 = a_0 x_2 = a_0 = x_1.$$

Hence we acquire a period-2 cycle and for all $n \geq 0$:

$$\begin{cases} x_{2n} = x_0 = 1, \\ x_{2n+1} = x_1 = a_0. \end{cases}$$

This is an example of a periodic cycle of a **Nonautonomous** Δ.**E.** that we will study in Chapters 2, 3, 4, and 5.

1.6 Complex Numbers and Periodic Cycles

We define a complex number in **rectangular coordinates** as:

$$z = x + yi,$$ (1.4)

where $Re(z) = x$, $Im(z) = y$, and $i = \sqrt{-1}$. In addition, $|z| = \sqrt{x^2 + y^2}$ and $\theta = arg(z)$ as shown in the diagram below (Figure 1.9):

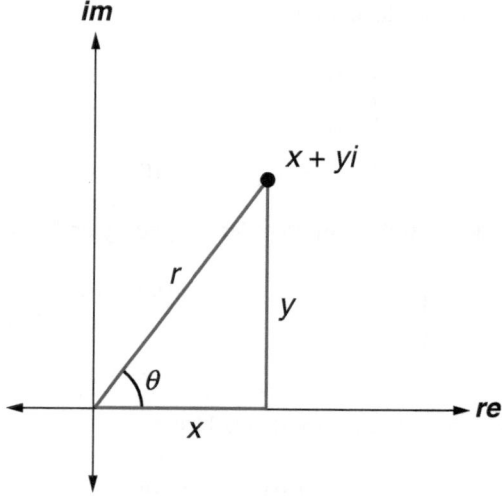

Fig. 1.9 Complex Number described in Rectangular Coordinates x and y. This diagram was designed by Olga A. Orlova (doctoral student at Munich Technical University).

Equation (1.4) is expressed in **rectangular coordinates** and we can write Equation (1.4) in **polar coordinates**:

$$z = |z|e^{i\theta} = |z|[cos(\theta) + isin(\theta)],$$

Our next goal is to examine periodicity of the following sequence of complex numbers:

$$\{z_n\}_{n=0}^{\infty} = (x + yi)^n = \left[|z|e^{i\theta}\right]^n = |z|^n e^{ni\theta} = |z|^n [cos(n\theta) + isin(n\theta)].$$

This is also called **De Moivre's Formula**. The following examples will illustrate various periods.

Example 18. The sequence of complex numbers:

$$\{z_n\}_{n=0}^{\infty} = (i)^n = \left[e^{i\left(\frac{\pi}{2}\right)}\right]^n = e^{ni\left(\frac{\pi}{2}\right)} = 1, i, -1, -i, \ldots,$$

is periodic with **period-4** and $\theta = arg(i) = \frac{\pi}{2}$ evoked by the following diagram (Figure 1.10):

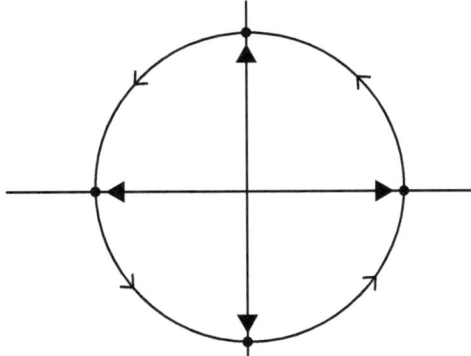

Fig. 1.10 Period-4 cycle with argument $\frac{\pi}{2}$. This diagram was designed by Olga A. Orlova (doctoral student at Munich Technical University).

Example 19. The sequence of complex numbers:

$$\{z_n\}_{n=0}^{\infty} = \left(-\frac{1}{2} + i\frac{\sqrt{3}}{2}\right)^n = \left[e^{i\left(\frac{2\pi}{3}\right)}\right]^n = e^{ni\left(\frac{2\pi}{3}\right)} = 1, \ -\frac{1}{2} + i\frac{\sqrt{3}}{2}, \ -\frac{1}{2} - i\frac{\sqrt{3}}{2}, \dots$$

is periodic with **period-3**; $\theta = arg\left(-\frac{1}{2} + i\frac{\sqrt{3}}{2}\right) = \frac{2\pi}{3}$ portrayed by the following diagram (Figure 1.11):

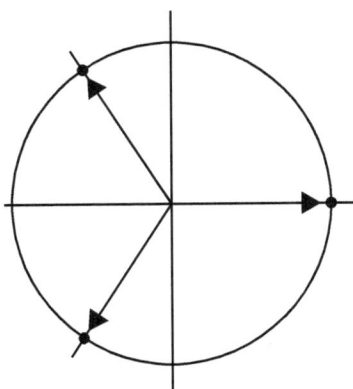

Fig. 1.11 Period-3 cycle with argument $\frac{2\pi}{3}$. This diagram was designed by Olga A. Orlova (doctoral student at Munich Technical University).

Therefore,

$$\{z_n\}_{n=0}^{\infty} = (x+yi)^n = |z|^n e^{ni\theta} \quad, \quad n = 0, 1, \dots,$$

is periodic with period $p \geq 3$ and expressed in the form:

$$z = e^{i\left(\frac{2\pi}{p}\right)} = \cos\left(\frac{2\pi}{p}\right) + \sin\left(\frac{2\pi}{p}\right)i.$$

Furthermore, it follows that:

(1) $|z| = \sqrt{x^2 + y^2} = 1$.

(2) $\arg(x+yi)$ is a rational multiple of π.

In Chapter 4 we will see the use of this idea when the characteristic polynomial

$$\lambda^2 + p\lambda + q = 0$$

has complex roots and the corresponding second order linear Δ.E. will exhibit periodic solutions corresponding to the argument of the complex roots.

1.7 Specific Patterns of Periodic Cycles

Our aim is to determine specific patterns of periodic cycles. In fact, how do we detect the pattern(s) when we move from one neighbor of the periodic cycle to the next? The next seven examples will manifest different patterns with shift of indices and change of negative signs.

Example 20. In the period-2 pattern:

$$\frac{1}{\beta+1} , \frac{-1}{\beta+1} , \dots,$$

the terms in the denominator stay the same from neighbor to neighbor while the negative sign switches only in the numerator. This is an example of an alternating period-2 cycle.

Example 21. In the period-2 pattern:

$$\frac{A_0 A_1 - 1}{1 + A_0} , \frac{A_0 A_1 - 1}{1 + A_1} , \dots.$$

From neighbor to neighbor the indices of the period-2 sequence $\{A_n\}_{n=0}^{\infty}$ in the numerator do not change, yet in the denominator the indices shift by 1 under modulo 2 arithmetic.

Example 22. In the period-2 pattern:

$$\frac{A_0 - A_1}{A_0 A_1 + 1}, \quad \frac{A_1 - A_0}{A_0 A_1 + 1}, \quad \cdots$$

From neighbor to neighbor the indices shift by 1 and the signs switch in the numerator of the period-2 sequence $\{A_n\}_{n=0}^{\infty}$ under modulo 2 arithmetic, but the indices do not change in the denominator. This is an example of an alternating period-2 cycle.

Example 23. In the period-4 pattern:

$$\frac{A_0}{A_0 A_1 + 1}, \quad \frac{A_1}{A_0 A_1 + 1}, \quad \frac{-A_0}{A_0 A_1 + 1}, \quad \frac{-A_1}{A_0 A_1 + 1}, \quad \cdots$$

From neighbor to neighbor the indices shift by 1 under modulo 2 arithmetic, the signs of the period-2 sequence $\{A_n\}_{n=0}^{\infty}$ switch in the numerator, and the sign alternates two terms later. However, the indices do not change in the denominator. This is an example of an alternating period-4 cycle.

Example 24. In the period-2 pattern:

$$\frac{A_1 B_0 + B_1}{1 + A_0 A_1}, \quad \frac{A_0 B_1 + B_0}{1 + A_0 A_1}, \quad \cdots,$$

the indices of the period-2 sequences $\{A_n\}_{n=0}^{\infty}$ and $\{B_n\}_{n=0}^{\infty}$ in the denominator do not change from neighbor to neighbor. However, the indices shift by 1 under modulo 2 arithmetic in the numerator.

Example 25. In the period-3 pattern:

$$\frac{\beta^2 + \beta - 1}{\beta^3 - 1}, \quad \frac{\beta^2 - \beta + 1}{\beta^3 - 1}, \quad \frac{-\beta^2 + \beta + 1}{\beta^3 - 1}, \quad \cdots,$$

the terms in the denominator stay the same from neighbor to neighbor but the negative sign switches in the numerator when moving between the terms $\beta^2, \beta, 1$.

Example 26. In the period-3 pattern:

$$\sqrt{\frac{A_0 A_1}{A_2}}, \quad \sqrt{\frac{A_1 A_2}{A_0}}, \quad \sqrt{\frac{A_2 A_0}{A_1}}, \quad \cdots$$

From neighbor to neighbor the indices of the period-3 sequence $\{A_n\}_{n=0}^{\infty}$ shift by 1 under modulo 3 arithmetic in both the numerator and the denominator.

More detailed examples of periodic patterns and alternating periodic patterns can be found in the **Appendix Chapter**.

1.8 Eventually Periodic Sequences (Solutions)

The sequence $\{x_n\}_{n=0}^{\infty}$ is **Eventually Periodic** with **minimal period p$\geq$ 2**, if there exists $N \geq 1$ such that

$$x_{n+N+p} = x_{n+N} \quad \text{for all} \quad n = 0, 1, \ldots .$$

In this case, we have N **transient terms**. We will illustrate three examples of eventually periodic solutions that exhibit transient terms or transient behavior.

Example 27. The sequence below is eventually periodic with period-2 with four transient terms:

$$[\, 32 \,,\, 16 \,,\, 8 \,,\, 4 \,] \,,\, 2 \,,\, 1 \,,\, 2 \,,\, 1 \,,\ldots ,$$

The four transient terms $x_0 - x_3$ are in square brackets and $x_6 = x_4$. This is an example of **3x+1 Conjecture** that we will meticulously study in Chapter 3.

Now we will inspect some graphical examples of eventually periodic solutions of various periods and with different number of transient terms.

Example 28. The graph below illustrates an eventually periodic solution with period-2 with six transient terms (Figure 1.12):

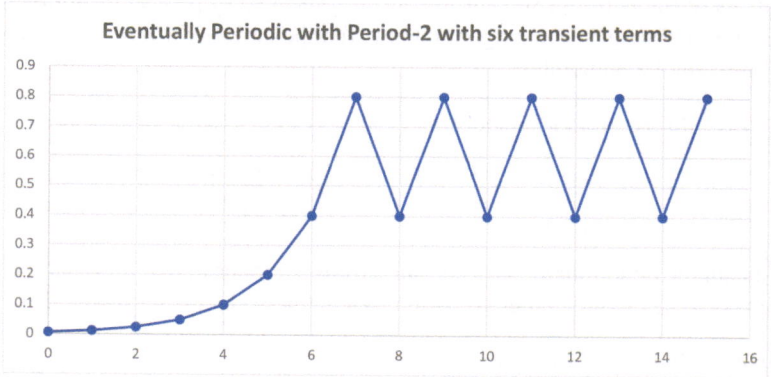

Fig. 1.12 Period-2 cycle with six increasing transient terms. Also $x_8 = x_6$. This is an example of **Tent-Map** that we will study in deeper details in Chapter 3.

Example 29. The graph below demonstrates an eventually periodic solution with period-4 with twelve transient terms (Figure 1.13):

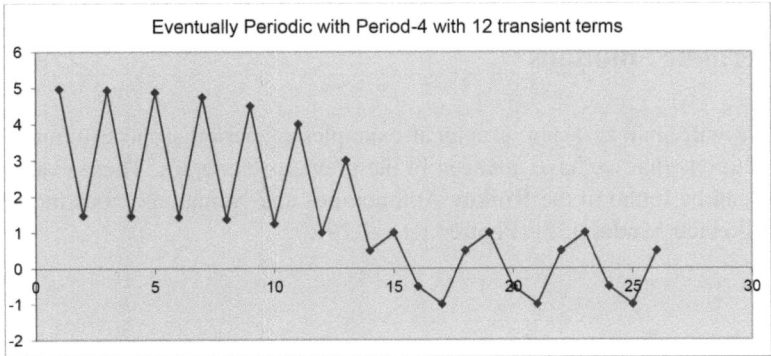

Fig. 1.13 Period-4 cycle with twelve transient terms that decrease in the pattern of two subsequences. In addition $x_{16} = x_{12}$. This is an example of a **piece-wise difference equation** that we will analyze in Chapter 3.

Example 30. The graph below demonstrates an eventually periodic solution with period-2 with six transient terms (Figure 1.14):

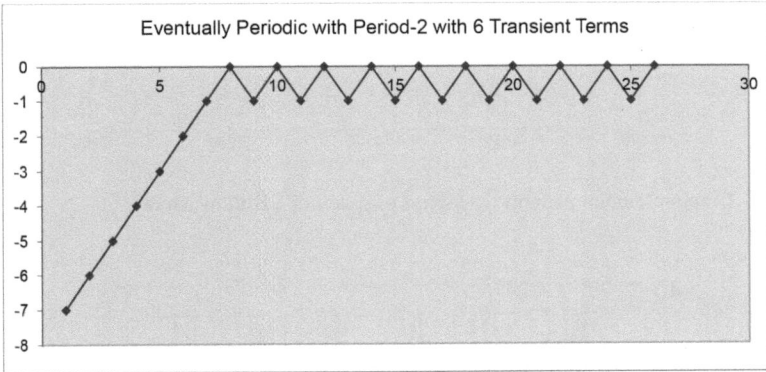

Fig. 1.14 Period-2 cycle with six transient terms that increase in a linear pattern. Also $x_9 = x_7$. This is an example of a **Piece-Wise Δ.E.** that we will analyze in Chapter 3.

More attributes to the existence of eventually periodic solutions will be exhibited on the Logistic Difference Equation and piece-wise difference equations in Chapter 3 and on Max-Type Difference Equations in Chapter 5.

1.9 Additional Examples of Periodic and Eventually Periodic Solutions

Now we will analyze a few graphical examples of periodic and eventually periodic solutions that we have not seen in the previous examples. These examples of graphs can be found in the **Rulkov** Autonomous and Nonautonomous models and the **Izhikevich Model**; [33] (Figures 1.15–1.19):

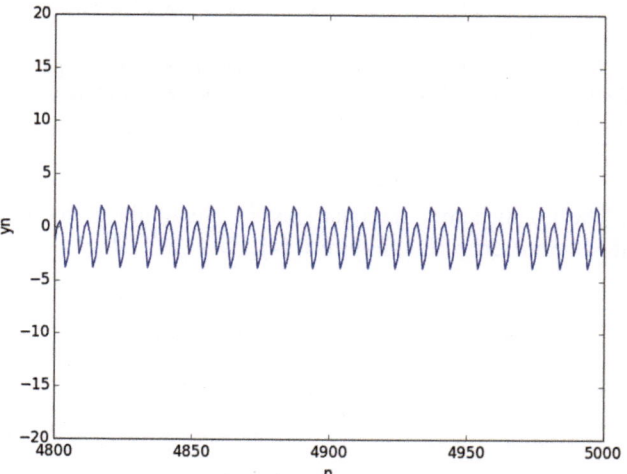

Fig. 1.15 Graph resembles patterns as systems of spikes of a **Rulkov Model**.

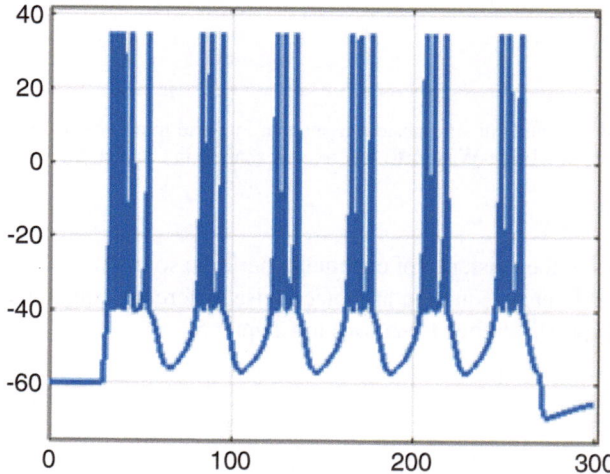

Fig. 1.16 Graph portrays clusters of spiking patterns of an **Izhikevich Model**. More details about these specific patterns can be found in [23].

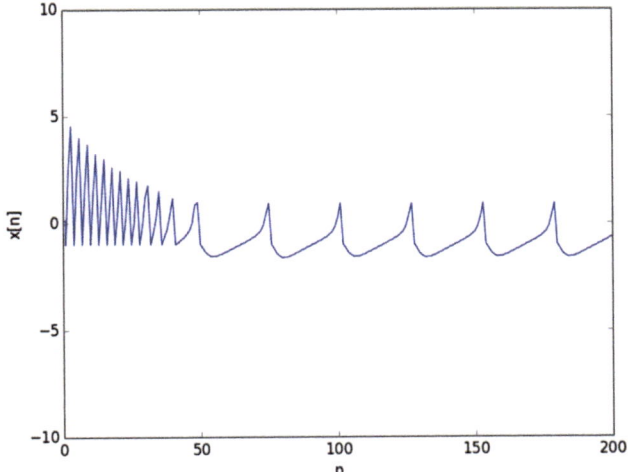

Fig. 1.17 Graph portrays eventually periodic solution with spiking patterns of a **Rulkov Model**. More details about these specific patterns can be found in [33].

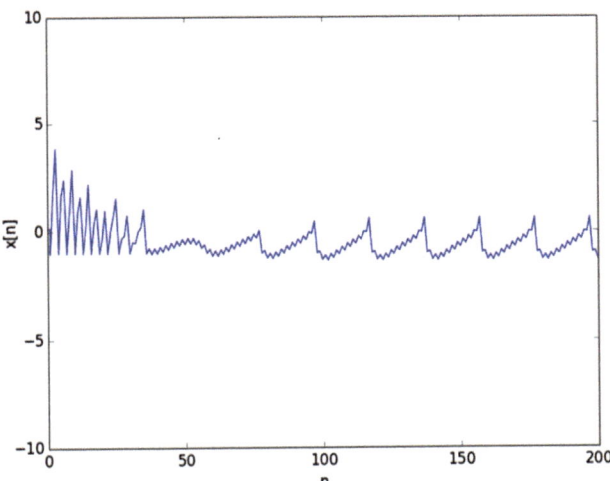

Fig. 1.18 Graph depicts much less regularity in the pattern of the transient terms and much more spiking patterns of periodic solutions of a **Rulkov Model**. More details about these specific patterns can be found in [33].

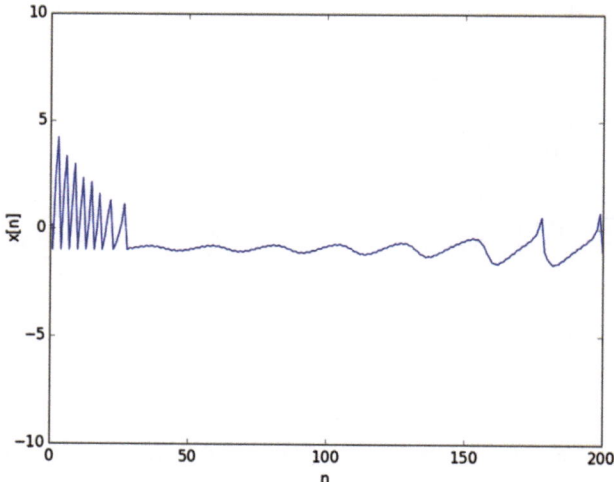

Fig. 1.19 Graph depicts chaotic behavior without any specific patterns of a **Rulkov Model**. More details about these specific patterns can be found in [33].

1.10 Exercises

In problems 1–6, write a **recursive formula (as an initial value problem)** for:

1:   12 , 17 , 22 , 27 , 32 , 37 , 42 ,... .

2:   7 , 19 , 31 , 43 , 55 , 67 , 79 ,... .

3:   3 , 5 , 9 , 15 , 23 , 33 , 45 ,... .

4:   1 , 4 , 9 , 16 , 25 , 36 , 49 ,... .

5:   4 , 7 , 16 , 31 , 52 , 79 , 112 ,... .

6:   9 , 13 , 21 , 33 , 49 , 69 , 93 ,... .

In problems 7–12, write a **recursive formula (as an initial value problem)** for:

7:   5 , 20 , 80 , 320 , 1280 , 5120 , ...

8:   54 , 18 , 6 , 2 , $\frac{2}{3}$, $\frac{2}{9}$, ...

9:   81 , 54 , 36 , 24 , 16 , $\frac{32}{3}$, ...

10: 1 , 2 , 8 , 64 , 1024 , ...

11: 1 , 3 , 15 , 105 , 945 , 10395 , ...

12: 1 , 5 , 45 , 585 , 9945 , ...

In problems 13–16, by iteration determine an **explicit solution** of:

13: $x_{n+1} = 4x_n$, $n = 0, 1, \ldots$.

14: $x_{n+1} = \frac{2x_n}{3}$, $n = 0, 1, \ldots$.

15: $x_{n+1} + 2x_n = 0$, $n = 0, 1, \ldots$.

16: $4x_{n+1} + 3x_n = 0$, $n = 0, 1, \ldots$.

In problems 17–25, determine the **period** of each initial value problem inductively:

17:
$$\begin{cases} x_{n+1} = \frac{x_n}{x_n - 1} \, , & n = 0, 1, \ldots , \\[2mm] x_0 = 5 \, . \end{cases}$$

18:
$$\begin{cases} x_{n+1} = \frac{x_n + 1}{x_n - 1} \, , & n = 0, 1, \ldots , \\[2mm] x_0 = 3 \, . \end{cases}$$

19:
$$\begin{cases} x_{n+1} = x_n + (-1)^n \, , & n = 0, 1, \ldots , \\[2mm] x_0 = 3 \, . \end{cases}$$

20:
$$\begin{cases} x_{n+1} = (-1)^n x_n + 2 \, , & n = 0, 1, \ldots , \\[2mm] x_0 = 8 \, . \end{cases}$$

21:
$$\begin{cases} x_{n+1} = \frac{(-1)^n}{x_n} \, , & n = 0, 1, \ldots , \\[2mm] x_0 = 4 \, . \end{cases}$$

22:
$$\begin{cases} x_{n+2} = x_{n+1} - x_n \, , & n = 0, 1, \ldots , \\[2mm] x_0 = 1 \, , \\[2mm] x_1 = 3 \, . \end{cases}$$

23:
$$\begin{cases} x_{n+2} = \sqrt{2}\, x_{n+1} - x_n \, , & n = 0, 1, \ldots , \\[2mm] x_0 = 3 \, , \\[2mm] x_1 = 5 \, . \end{cases}$$

24:
$$\begin{cases} x_{n+2} = \frac{1}{x_n}, & n = 0, 1, \ldots, \\ x_0 = 2, \\ x_1 = 4. \end{cases}$$

25:
$$\begin{cases} x_{n+2} = \frac{x_{n+1}}{x_n}, & n = 0, 1, \ldots, \\ x_0 = 2, \\ x_1 = 6. \end{cases}$$

In problems 26–32, determine if each given sequence is **periodic**. If not, explain why. If so, determine the period.

26. $x_n = \left(\frac{1}{\sqrt{2}} + \frac{1}{\sqrt{2}} i \right)^n$, $n = 0, 1, \ldots$.

27. $x_n = \left(\frac{1}{2} + \frac{\sqrt{3}}{2} i \right)^n$, $n = 0, 1, \ldots$.

28. $x_n = \left(\frac{\sqrt{3}}{2} + \frac{1}{2} i \right)^n$, $n = 0, 1, \ldots$.

29. $x_n = (-i)^n$, $n = 0, 1, \ldots$.

30. $x_n = \left(\frac{i}{2} \right)^n$, $n = 0, 1, \ldots$.

31. $x_n = \left(\sqrt{2} + \sqrt{2} i \right)^n$, $n = 0, 1, \ldots$.

32. $x_n = \left(\frac{4}{5} + \frac{3}{5} i \right)^n$, $n = 0, 1, \ldots$.

In problems 33–38, graphically determine the period:

33.

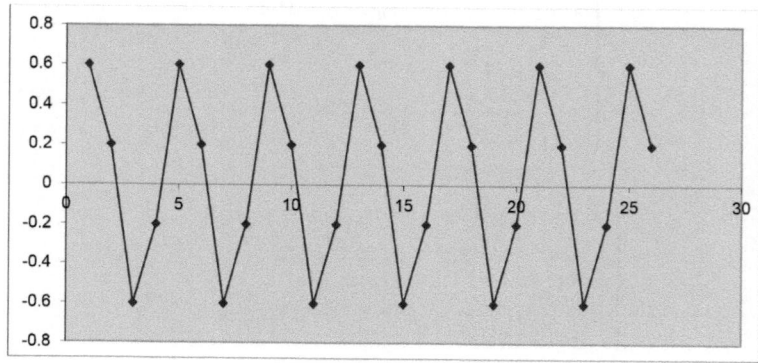

34.

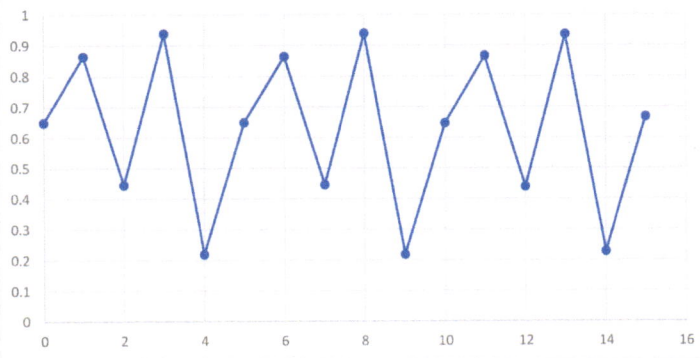

35.

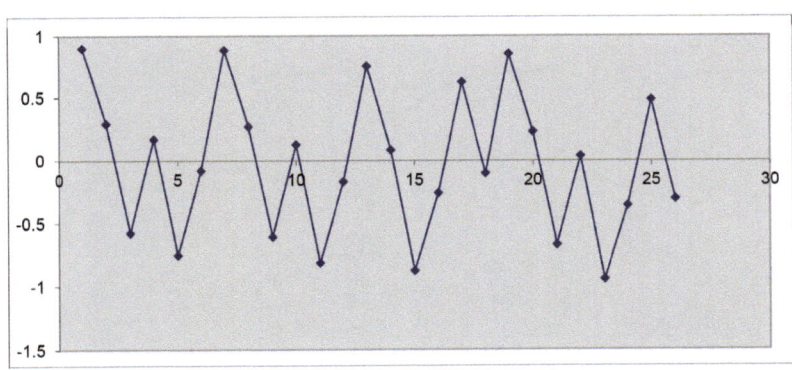

36.

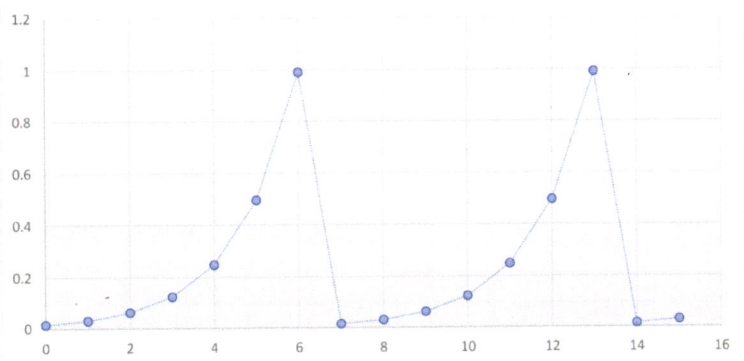

37.

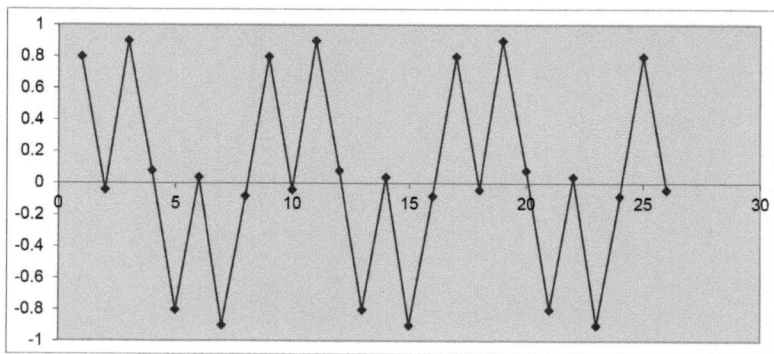

38.

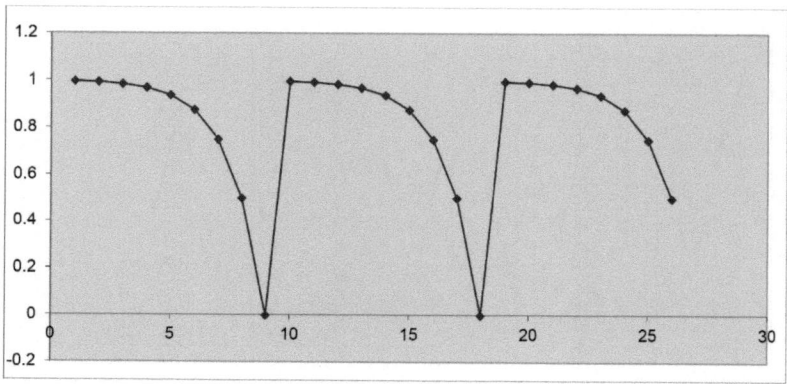

In problems 39–49, as in Examples (20)–(26), describe the pattern of the given periodic cycle:

39. Describe the pattern of the period-3 cycle:

$$\frac{2}{9}, \frac{4}{9}, \frac{8}{9}, \dots .$$

40. Suppose that $\{a_n\}_{n=0}^{\infty}$ is period-3 sequence. Describe the pattern of the period-3 cycle:

$$\frac{a_0 - a_1 + a_2}{2}, \frac{a_0 + a_1 - a_2}{2}, \frac{-a_0 + a_1 + a_2}{2}, \dots .$$

41. Suppose that $\{a_n\}_{n=0}^{\infty}$ is period-4 sequence. Describe the pattern of the period-4 cycle:

$$\frac{a_0 + \sqrt{2}a_1 + a_2}{2}, \frac{a_1 + \sqrt{2}a_2 + a_3}{2}, \frac{a_2 + \sqrt{2}a_3 + a_0}{2}, \frac{a_3 + \sqrt{2}a_0 + a_1}{2}, \dots .$$

42. Suppose that $\{A_n\}_{n=0}^{\infty}$ is period-4 sequence. Describe the pattern of the period-4 cycle:

$$0\,,\ \frac{A_0A_2-1}{1+A_0}\,,\ 0\,,\ \frac{A_0A_2-1}{1+A_2}\,,\ \ldots.$$

43. Suppose that $\{A_n\}_{n=0}^{\infty}$ is period-4 sequence. Describe the pattern of the period-4 cycle:

$$\frac{A_1A_3-1}{1+A_1}\,,\ 0\,,\ \frac{A_1A_3-1}{1+A_3}\,,\ 0\,,\,,\ \ldots.$$

44. Suppose that $\{a_n\}_{n=0}^{\infty}$ is period-4 sequence. Describe the pattern of the period-4 cycle:

$$\frac{a_0a_1a_2a_3-1}{a_0a_1a_2-a_0a_1+a_0-1}\,,\quad \frac{a_0a_1a_2a_3-1}{a_1a_2a_3-a_1a_2+a_1-1}\,,$$

$$\frac{a_0a_1a_2a_3-1}{a_2a_3a_0-a_2a_3+a_2-1}\,,\quad \frac{a_0a_1a_2a_3-1}{a_3a_0a_1-a_3a_0+a_3-1}\,,\ \ldots.$$

45. Suppose that $\{A_n\}_{n=0}^{\infty}$ and $\{B_n\}_{n=0}^{\infty}$ are period-2 sequences. Describe the pattern of the period-4 cycle:

$$\frac{A_1-B_0}{2}\,,\ \frac{B_1-A_0}{2}\,,\ \frac{B_1+A_0}{2}\,,\ \frac{A_1+B_0}{2}\,,\ \ldots.$$

46. Suppose that $\{A_n\}_{n=0}^{\infty}$ is period-6 sequence. Describe the pattern of the period-6 cycle:

$$\sqrt{\frac{A_4A_0}{A_2}}\,,\ \sqrt{\frac{A_5A_1}{A_3}}\,,\ \sqrt{\frac{A_0A_2}{A_4}}\,,\ \sqrt{\frac{A_1A_3}{A_5}}\,,\ \sqrt{\frac{A_2A_4}{A_0}}\,,\ \sqrt{\frac{A_3A_5}{A_1}}\,,\ \ldots.$$

47. Suppose that $\{A_n\}_{n=0}^{\infty}$ is period-3 sequence. Describe the pattern of the period-6 cycle:

$$\frac{A_0A_1A_2-1}{1+A_0+A_0A_2}\,,\ 0\,,\ \frac{A_0A_1A_2-1}{1+A_1+A_1A_0}\,,\ 0\,,\ \frac{A_0A_1A_2-1}{1+A_2+A_2A_1}\,,\ 0\,\ldots.$$

48. Suppose that $\{A_n\}_{n=0}^{\infty}$ is period-5 sequence. Describe the pattern of the period-10 cycle:

$$1\,,\ 0\,,\ A_0\,,\ 0\,,\ A_2A_0\,,\ 0\,,\ A_4A_2A_0\,,\ 0\,,\ A_1A_4A_2A_0\,,\ 0\,,\ldots.$$

49. Suppose that $\{A_n\}_{n=0}^{\infty}$ is period-8 sequence. Describe the pattern of the period-8 cycle:

$$0, \ \frac{A_1A_3A_5A_7 - 1}{1 + A_1 + A_1A_3 + A_1A_3A_5}, \ 0, \ \frac{A_1A_3A_5A_7 - 1}{1 + A_3 + A_3A_5 + A_3A_5A_7},$$

$$0, \ \frac{A_1A_3A_5A_7 - 1}{1 + A_5 + A_5A_7 + A_5A_7A_1}, \ 0, \ \frac{A_1A_3A_5A_7 - 1}{1 + A_7 + A_7A_1 + A_7A_1A_3}, \ \ldots$$

Chapter 2
First Order Linear Difference Equations and Patterns of Periodic Solutions

2.1 First Order Linear Difference Equations

Our goal of this chapter is to understand how to solve first order linear difference equations explicitly (by induction) and to study the periodic traits and patterns of first order linear difference equations. In addition, we will break up this chapter into three sections: Periodic Solutions of First Order Linear Homogeneous Difference Equations, First Order Nonhomogeneous Difference Equations, and First Order Nonautonomous Linear Difference Equations. Here are three examples of first order linear difference equations:

(i) (Homogeneous) $x_{n+1} = \frac{2x_n}{3}$, $n = 0, 1, \ldots$.

(ii) (Nonhomogeneous) $x_{n+1} = 4x_n - 1$, $n = 0, 1, \ldots$.

(iii) (Nonautonomous) $x_{n+1} = x_n - \left(\frac{-1}{2}\right)^n$, $n = 0, 1, \ldots$.

Our aim is to solve each linear Δ.E. explicitly, check the solution, and analyze the long-term behavior of the solutions and the periodic traits of the solutions. In addition, our intent of this chapter is to explicitly solve the linear Δ.E. in the form:

$$x_{n+1} = a_n x_n + b_n , \quad n = 0, 1, \ldots ,$$

where $\{a_n\}_{n=0}^{\infty}$ and $\{b_n\}_{n=0}^{\infty}$ are periodic sequences with either the same period or with different periods and $x_0 \in \Re$. First, we will begin our investigation with the **Homogeneous Δ.E.** in the form:

$$x_{n+1} = a x_n , \quad n = 0, 1, \ldots ,$$

where $a \neq 0$ and $x_0 \in \Re$.

© Springer Nature Switzerland AG 2018
M. A. Radin, *Periodic Character and Patterns of Recursive Sequences*,
https://doi.org/10.1007/978-3-030-01780-4_2

2.2 Homogeneous First Order Linear Difference Equations

The following are examples of linear homogeneous difference equations of first order:

(i) $x_{n+1} = 4x_n$, $n = 0, 1, \ldots$.

(ii) $x_{n+1} = \frac{3x_n}{4}$, $n = 0, 1, \ldots$.

(iii) $x_{n+1} = -x_n$, $n = 0, 1, \ldots$.

In (i), the coefficient $a > 1$. In (ii), the coefficient $0 < a < 1$. In (iii), the coefficient $a = -1$. In this section, we begin our examination of the first-order Δ.E. in the form:

$$x_{n+1} - ax_n = 0 \ , \quad n = 0, 1, \ldots , \tag{2.1}$$

where the parameter $a \neq 0$ and the initial condition $x_0 \in \mathfrak{R}$. Let $\{x_n\}_{n=0}^{\infty}$ be a solution of Equation (2.1). Then it follows by iteration and induction that

$$x_n = a^n x_0 \ , \quad n = 0, 1, \ldots . \tag{2.2}$$

Furthermore, notice that $\bar{x} = 0$ is the only equilibrium point of Equation (2.1) provided that $a \neq 1$. In addition, if $a = 1$, then every initial condition x_0 is an equilibrium point of Equation (2.1). Now we will break up into the following cases. In fact, observe that:

(i)   When $|a| < 1$, then

$$\lim_{n \to \infty} x_n = \lim_{n \to \infty} a^n x_0 = 0 .$$

(ii) When $a = 1$, then

$$\lim_{n \to \infty} x_n = \lim_{n \to \infty} x_0 = x_0 .$$

(iii) When $a = -1$, then

$$x_n = (-1)^n x_0 = \begin{cases} x_0 & \text{if n is even,} \\ \\ -x_0 & \text{if n is odd.} \end{cases}$$

In this case, we see that every nontrivial (nonequilibrium solution) solution of Equation (2.1) is periodic with period-2.

(iv) When $a > 1$, then

$$\lim_{n \to \infty} x_n = \lim_{n \to \infty} a^n x_0 = \begin{cases} \infty & \text{if } x_0 > 0, \\ \\ -\infty & \text{if } x_0 < 0. \end{cases}$$

(v) When $a < -1$, then we get the following two cases:

–   If $x_0 > 0$, then

$$\lim_{n \to \infty} x_{2n} = +\infty \quad \text{and} \quad \lim_{n \to \infty} x_{2n+1} = -\infty .$$

–   If $x_0 < 0$, then

$$\lim_{n \to \infty} x_{2n} = -\infty \quad \text{and} \quad \lim_{n \to \infty} x_{2n+1} = +\infty .$$

From (iii), when $a = -1$, then every nontrivial (nonequilibrium solution) solution of Equation (2.1) is **oscillatory** and periodic with the following alternating period-2 pattern:

$$x_0 , \; -x_0 , \; x_0 , \; -x_0, \; \dots .$$

Furthermore, notice that the sum of two neighboring terms is always 0. Moreover, this is a frequent phenomena that will appear in other situations when periodic cycles exist. We will explore deeper analysis of periodicity properties in the section **Nonhomogeneous Case**. In the next three examples, we will go through the details of solving the homogeneous first order linear difference equations and illustrate the monotonic behavior of solutions graphically (Figures 2.1–2.3).

Example 1. Solve the Initial Value Problem:

$$\begin{cases} x_{n+1} = \frac{x_n}{2} , & n = 0, 1, \dots, \\ \\ x_0 = \frac{1}{2} , \end{cases}$$

and:

(i)   Determine

$$\lim_{n \to \infty} x_n .$$

(ii) Sketch the graph and describe the monotonic character.

Solution: Observe that from Equations (2.1) and (2.2), we get

$$x_n = x_0 \left(\frac{1}{2} \right)^n = \frac{1}{2} \left(\frac{1}{2} \right)^n = \left(\frac{1}{2} \right)^{n+1} .$$

Hence:

$$\lim_{n \to \infty} x_n = \lim_{n \to \infty} \left(\frac{1}{2} \right)^{n+1} = 0 .$$

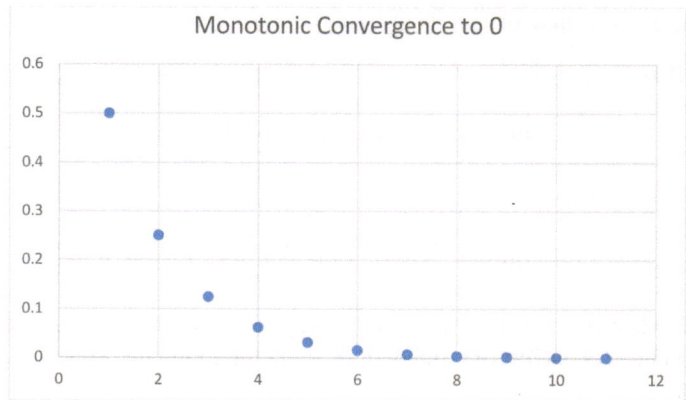

Fig. 2.1 The solution monotonically approaches zero from above. We obtain the monotonic convergence $x_{n+1} < x_n$ for all $n \geq 0$ when the coefficient $a > 0$.

Example 2. Solve the Initial Value Problem:

$$\begin{cases} x_{n+1} = \frac{-4x_n}{5} \; , \quad n = 0, 1, \ldots, \\ x_0 = 1 \; , \end{cases}$$

and:

(i) Determine

$$\lim_{n \to \infty} x_n \; .$$

(ii) Sketch the graph and describe the monotonic character.

Solution: Observe that from Equations (2.1) and (2.2), we get

$$x_n = x_0 \left(\frac{-4}{5} \right)^n = \left(\frac{-4}{5} \right)^n = (-1)^n \left(\frac{4}{5} \right)^n .$$

Thus:

$$\lim_{n \to \infty} x_n = (-1)^n \left(\frac{4}{5} \right)^n = 0 .$$

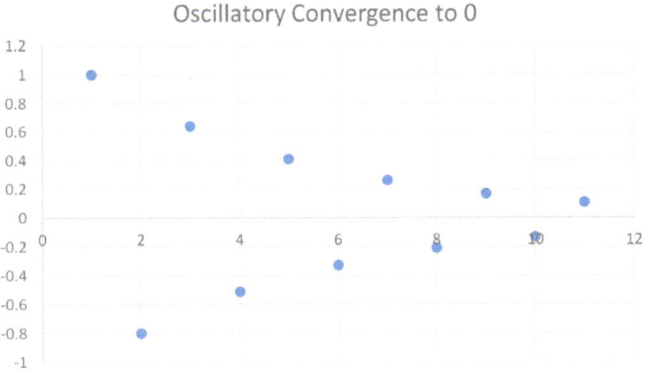

Fig. 2.2 The solution converges to 0 in two subsequences; one subsequence from above and one subsequence from below; $x_{2n+2} < x_{2n}$ and $x_{2n+1} < x_{2n+3}$ for all $n \geq 0$. This oscillatory behavior occurs when the coefficient $a < 0$.

Example 3. Solve the Initial Value Problem:

$$\begin{cases} x_{n+1} = \frac{3x_n}{2} \ , & n = 0, 1, \dots, \\ x_0 = \frac{9}{4} \ , \end{cases}$$

and:

(i) Determine

$$\lim_{n \to \infty} x_n \ .$$

(ii) Sketch the graph and describe the monotonic character.

Solution: Observe that from Equations (2.1) and (2.2), we get

$$x_n = x_0 \left(\frac{3}{2}\right)^n = \frac{9}{4}\left(\frac{3}{2}\right)^n = \left(\frac{3}{2}\right)^{n+2} \ .$$

Therefore:

$$\lim_{n \to \infty} x_n = \lim_{n \to \infty} \left(\frac{3}{2}\right)^{n+2} = +\infty \ .$$

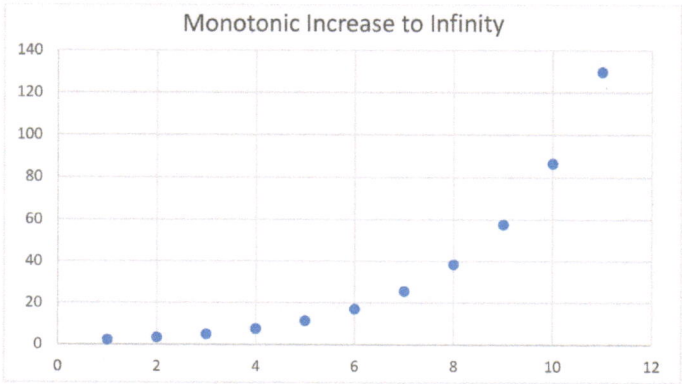

Fig. 2.3 The solution monotonically diverges to $+\infty$ and we see that $x_n < x_{n+1}$ for all $n \geq 0$.

In the next two examples we will work out the details of solving the given Δ.E. and the given **Initial Value Problem** analytically.

Example 4. Solve the Initial Value Problem:

$$\begin{cases} 4x_{n+1} - 3x_n = 0 \ , \quad n = 0, 1, \ldots, \\[2mm] x_0 = \tfrac{4}{3} \, , \end{cases}$$

and:

(i)　Determine

$$\lim_{n \to \infty} x_n \, .$$

(ii)　Verify that the solution is correct.

Solution: From Equations (2.1) and (2.2), we get

$$x_n = x_0 \left(\frac{3}{4} \right)^n = \frac{4}{3} \left(\frac{3}{4} \right)^n = \left(\frac{3}{4} \right)^{n-1} \, .$$

Now we see that

$$\lim_{n \to \infty} x_n = \lim_{n \to \infty} \left(\frac{3}{4} \right)^{n-1} = 0 \, .$$

Furthermore:

$$4x_{n+1} - 3x_n = 4 \left(\frac{3}{4} \right)^n - 3 \left(\frac{3}{4} \right)^{n-1} = \frac{3^n}{4^{n-1}} - \frac{3^n}{4^{n-1}} = 0 \, .$$

Example 5. Solve the Initial Value Problem:

$$\begin{cases} 4x_{n+1} + 6x_n = 0 \ , \quad n = 0, 1, \ldots, \\[2mm] x_0 = 1 \, , \end{cases}$$

and:

(i) Determine

$$\lim_{n \to \infty} x_n .$$

(ii) Verify that the solution is correct.

Solution: From Equations (2.1) and (2.2), we get

$$x_n = \left(-\frac{3}{2}\right)^n = (-1)^n \left(\frac{3}{2}\right)^n .$$

Then we see that

$$\lim_{n \to \infty} x_n = \lim_{n \to \infty} (-1)^n \left(\frac{3}{2}\right)^n$$

does not exists as

$$\lim_{n \to \infty} x_{2n} = +\infty \quad \text{and} \quad \lim_{n \to \infty} x_{2n+1} = -\infty .$$

Hence:

$$4x_{n+1} + 6x_n = 4\left(\frac{-3}{2}\right)^{n+1} + 6\left(\frac{-3}{2}\right)^n = -6(-1)^n \left(\frac{3}{2}\right)^n + 6(-1)^n \left(\frac{3}{2}\right)^n = 0.$$

Now we will proceed with examining the **Linear Nonhomogeneous** Δ.**E.** in the form:

$$x_{n+1} = ax_n + b \quad , \quad n = 0, 1, \dots ,$$

where $a \neq 0$, $b \neq 0$, and $x_0 \in \Re$.

2.3 Nonhomogeneous First Order Linear Difference Equations

The following are examples of first order linear nonhomogeneous difference equations:

(i) $x_{n+1} = 3x_n - 5$, $n = 0, 1, \dots$.

(ii) $x_{n+1} = \frac{x_n}{3} - 4$, $n = 0, 1, \dots$.

In this section we will investigate the **Nonhomogeneous** Δ.**E.** in the form:

$$x_{n+1} - ax_n = b \quad , \quad n = 0, 1, \dots , \tag{2.3}$$

where $a \neq 0$, $b \neq 0$, and $x_0 \in \Re$. Notice that $\bar{x} = \frac{b}{1-a}$ is the only equilibrium point of Equation (2.3) provided that $a \neq 1$. By solving Equation (2.3) explicitly by iteration and induction, we get the following two cases:

(i) When $a = 1$,

$$x_n = x_0 + nb . \tag{2.4}$$

In this case we see that Equation (2.3) has no equilibrium point.

(ii) When $a = -1$, then

$$x_n = \begin{cases} x_0 & \text{if n is even,} \\[2ex] -x_0 + b & \text{if n is odd.} \end{cases}$$

In this case, every nontrivial solution of Equation (2.3) is periodic with period-2.

(iii) In all other cases, it follows that

$$x_n = a^n x_0 + b(1 + a + a^2 + \ldots + a^{n-1}) = a^n x_0 + \frac{b(1 - a^n)}{1 - a},$$

$$= a^n \left(x_0 - \frac{b}{1-a} \right) + \frac{b}{1-a},$$

which can be rewritten as

$$x_n = a^n (x_0 - \bar{x}) + \bar{x}. \tag{2.5}$$

From (ii), when $a = -1$, then every nontrivial (nonequilibrium solution) solution of Equation (2.3) is **oscillatory** and periodic with the following alternating period-2 pattern:

$$x_0, \ -x_0 + b, \ x_0, \ -x_0 + b, \ \ldots .$$

Moreover, the sum of two neighboring terms is always b. We will examine similar periodicity properties in the **Nonautonomous Case**. In the next three examples, we will solve the nonhomogeneous first order linear Δ.E. analytically and verify the solution.

Example 6. Solve the Initial Value Problem:

$$\begin{cases} 2x_{n+1} - x_n = 4, \quad n = 0, 1, \ldots, \\[2ex] x_0 = 8, \end{cases}$$

and determine

$$\lim_{n \to \infty} x_n .$$

Solution: From Equation (2.3), we get

$$x_n = \left(\frac{1}{2} \right)^n [8 - 4] + 4 = \left(\frac{1}{2} \right)^{n-2} + 4 .$$

Thus

$$\lim_{n \to \infty} x_n = \lim_{n \to \infty} \left(\frac{1}{2} \right)^{n-2} + 4 = 4 .$$

Example 7. Solve the Initial Value Problem:

$$\begin{cases} 2x_{n+1} + 6x_n = -3 , & n = 0,1,\ldots, \\[2mm] x_0 = -\frac{1}{2} , \end{cases}$$

and:

(i) Determine

$$\lim_{n \to \infty} x_n .$$

(ii) Verify that the solution is correct.

Solution: Notice that from Equations (2.3) and (2.5), we get

$$x_n = (-3)^n \left(-\frac{1}{2} - \frac{3}{8} \right) - \frac{3}{8} = -\frac{(-3)^n}{8} - \frac{3}{8} .$$

Now we see that

$$\lim_{n \to \infty} x_n = \lim_{n \to \infty} \left(-\frac{(-3)^n}{8} - \frac{3}{8} \right) ,$$

does not exists as

$$\lim_{n \to \infty} x_{2n} = -\infty \quad \text{and} \quad \lim_{n \to \infty} x_{2n+1} = +\infty .$$

Furthermore, it follows that

$$2x_{n+1} + 6x_n = 2\left[-\frac{(-3)^{n+1}}{8} - \frac{3}{8} \right] + 6\left[-\frac{(-3)^n}{8} - \frac{3}{8} \right]$$

$$= \frac{3}{4}(-3)^n - \frac{3}{4} + -\frac{3}{4}(-3)^n - \frac{9}{4} = -3 .$$

Example 8. Solve the Initial Value Problem:

$$\begin{cases} 3x_{n+1} - 3x_n = -2 , & n = 0,1,\ldots, \\[2mm] x_0 = -3 , \end{cases}$$

and determine

$$\lim_{n \to \infty} x_n .$$

Solution: Observe that from Equations (2.3) and (2.4), we get

$$x_n = -3 - \frac{2}{3}n$$

Thus, we see that

$$\lim_{n \to \infty} x_n = \lim_{n \to \infty} \left(-3 - \frac{2}{3}n \right) = -\infty .$$

Now we will move on with studying the Nonautonomous Δ.E. in the form:

$$x_{n+1} = a_n x_n + b_n \quad , \quad n = 0, 1, \ldots ,$$

where $\{a_n\}_{n=0}^{\infty}$ and $\{b_n\}_{n=0}^{\infty}$ are periodic sequences with either the same period or with different periods and $x_0 \in \mathfrak{R}$.

2.4 Nonautonomous First Order Linear Difference Equations

We will commence with examples of first order linear nonautonomous difference equations:

(i) $x_{n+1} = x_n + \frac{1}{n^2} \quad , \quad n = 0, 1, \ldots .$

(ii) $x_{n+1} = x_n + \frac{n}{2^n} \quad , \quad n = 0, 1, \ldots .$

(iii) $x_{n+1} = 2^n x_n + \frac{1}{4^n} \quad , \quad n = 0, 1, \ldots .$

Our goal is to investigate the periodic traits of Nonautonomous First Order Linear Difference Equations and will encounter existence of unique periodic cycles in some instances. We will emerge with the periodic traits of the first order homogeneous nonautonomous Δ.E.:

$$x_{n+1} = a_n x_n \quad , \quad n = 0, 1, \ldots , \tag{2.6}$$

where $x_0 \neq 0$ and $\{a_n\}_{n=0}^{\infty}$ is a periodic period-k sequence ($k \geq 2$). The next two examples will carefully examine the details in determining the necessary and sufficient conditions for the existence of periodic cycles and their patterns of Equation (2.6).

Example 9. Suppose that $\{a_n\}_{n=0}^{\infty}$ is a periodic period-2 sequence. Determine the necessary and sufficient conditions for the existence of period-2 cycles of:

$$x_{n+1} = a_n x_n \quad , \quad n = 0, 1, \ldots .$$

Solution: By iteration we get:

$$x_0 ,$$
$$x_1 = a_0 x_0 ,$$
$$x_2 = a_1 x_1 = a_1 [a_0 x_0] = x_0 .$$

Note that $x_2 = x_0$ if and only if $a_0 a_1 = 1$ with the following period-2 pattern:

$$x_0 , a_0 x_0 , x_0 , a_0 x_0 , \ldots .$$

Unless $a_0a_1 = 1$, the existence of period-2 solutions is not possible. We will see similar conditions for the existence of periodic cycles in future examples. Also when $a_0a_1 = -1$, then Equation (2.6) will have alternating period-4 cycles.

Example 10. Suppose that $\{a_n\}_{n=0}^{\infty}$ is a periodic period-3 sequence. Determine the necessary and sufficient conditions for the existence of period-3 cycles of:

$$x_{n+1} = a_nx_n \ , \quad n = 0, 1, \ldots .$$

Solution: By iteration we get:

$$
\begin{aligned}
&x_0 \ , \\
&x_1 = a_0x_0 \ , \\
&x_2 = a_1x_1 = a_1\left[a_0x_0\right] , \\
&x_3 = a_2x_2 = a_2\left[a_0a_1x_0\right] = x_0 \ .
\end{aligned}
$$

Note that $x_3 = x_0$ if and only if $a_0a_1a_2 = 1$ with the following period-3 pattern:

$$x_0 \ , \ a_0x_0 \ , \ a_0a_1x_0 \ , \ \ldots .$$

Unless $a_0a_1a_2 = 1$, the existence of period-3 solutions is not possible. When $a_0a_1a_2 = -1$, then Equation (2.6) will have alternating period-6 cycles.

From Examples 9 and 10, the following theorems describe the conclusions.

Theorem 1. *Suppose that $\{a_n\}_{n=0}^{\infty}$ is a period-k sequence, $(k \geq 2)$. Then every solution of:*

$$x_{n+1} = a_nx_n \ , \quad n = 0, 1, \ldots ,$$

is periodic with period-k if and only if:

$$\prod_{i=1}^{k} a_i = 1 .$$

Proof: *Similar to Example 10, we obtain:*

$$x_0 \ ,$$

$$x_1 = a_0 \, x_0 \ ,$$

$$x_2 = a_1x_1 = \left[a_1a_0\right] x_0 \ ,$$

$$x_3 = a_2x_2 = \left[a_2a_1a_0\right] x_0 \ ,$$

$$x_4 = a_3x_2 = \left[a_3a_2a_1a_0\right] x_0 \ ,$$

$$x_5 = a_4x_3 = \left[a_4a_3a_2a_1a_0\right] x_0 \ ,$$

$$\vdots$$

By induction for all $n \geq 0$:

$$x_{n+1} = \left[\prod_{i=0}^{n} a_i \right] x_0 .$$

Theorem 2. *Suppose that $\{a_n\}_{n=0}^{\infty}$ is a period-k sequence, ($k \geq 2$). Then every solution of:*

$$x_{n+1} = -a_n x_n \quad , \quad n = 0, 1, \dots ,$$

is periodic with period-2k if and only if:

$$\prod_{i=1}^{k} a_i = -1 .$$

Proof: *The proof is similar to the proof of Theorem 1 and will be omitted.*

We will shift our focus on the periodic traits of:

$$x_{n+1} = x_n + a_n \quad , \quad n = 0, 1, \dots , \tag{2.7}$$

where $x_0 \in \Re$ and $\{a_n\}_{n=0}^{\infty}$ is a periodic period-k sequence ($k \geq 2$). The next two examples will analyze the details for the existence of periodic cycles and their patterns of Equation (2.7).

Example 11. Suppose that $\{a_n\}_{n=0}^{\infty}$ is a periodic period-2 sequence. Determine the necessary and sufficient conditions for the existence of period-2 cycles of:

$$x_{n+1} = x_n + a_n \quad , \quad n = 0, 1, \dots .$$

Solution: By iteration we get:

$$x_0 ,$$

$$x_1 = x_0 + a_0 ,$$

$$x_2 = x_1 + a_1 = [x_0 + a_0] + a_1 = x_0$$

$$x_3 = x_2 + a_0 = x_0 + a_0 = x_1 .$$

Hence $x_2 = x_0$ if and only if $a_0 + a_1 = 0$ with the following alternating period-2 pattern:

$$x_0 , x_0 + a_0 , x_0 , x_0 + a_0 , \dots .$$

Example 12. Suppose that $\{a_n\}_{n=0}^{\infty}$ is a period-3 sequence. Determine the necessary and sufficient conditions for the existence of period-3 cycles of:

$$x_{n+1} = x_n + a_n \quad , \quad n = 0, 1, \dots .$$

Solution: Notice:

$$x_0 ,$$

$$x_1 = x_0 + a_0 ,$$

$$x_2 = x_1 + a_1 = [x_0 + a_0] + a_1 ,$$

$$x_3 = x_2 + a_2 = [x_0 + a_0 + a_1] + a_2 = x_0 .$$

Hence $x_3 = x_0$ if and only if $a_0 + a_1 + a_2 = 0$ with the following period-3 pattern:

$$x_0 , x_0 + a_0 , x_0 + a_0 + a_1 , x_0 , x_0 + a_0 , x_0 + a_0 + a_1 , \dots .$$

From Examples 11 and 12 the following theorem outlines the conclusions.

Theorem 3. *Suppose that $\{a_n\}_{n=0}^{\infty}$ is a period-k sequence, $(k \geq 2)$. Then every solution of:*

$$x_{n+1} = x_n + a_n , \quad n = 0, 1, \dots ,$$

is periodic with period-k if and only if:

$$\sum_{i=0}^{k-1} a_i = 0 .$$

Proof: *Similar to Example 12, we acquire:*

$$x_0 ,$$

$$x_1 = x_0 + a_0 ,$$

$$x_2 = x_1 + a_1 = x_0 + a_0 + a_1 ,$$

$$x_3 = x_2 + a_2 = x_0 + a_0 + a_1 + a_2 ,$$

$$x_4 = x_3 + a_3 = x_0 + a_0 + a_1 + a_2 + a_3 ,$$

$$x_5 = x_4 + a_4 = x_0 + a_0 + a_1 + a_2 + a_3 + a_4 ,$$

$$\vdots$$

Then for all $n \in \mathbb{N}$:

$$x_n = x_0 + \left[\sum_{i=0}^{n-1} a_i \right] .$$

The following example will examine the periodic nature of solutions of the following nonautonomous Δ.E.:

$$x_{n+1} = a_n x_n + 1 \quad, \quad n = 0, 1, \ldots ,$$

where $\{a_n\}_{n=0}^{\infty}$ is a period-k sequence, $(k \geq 2)$.

Example 13. Suppose that $\{a_n\}_{n=0}^{\infty}$ is a period-2 sequence. Determine the necessary and sufficient conditions for the existence of period-4 cycles of:

$$x_{n+1} = a_n x_n + 1 \quad, \quad n = 0, 1, \ldots .$$

Solution: Observe that:

$$x_0 ,$$

$$x_1 = a_0 x_0 + 1$$

$$x_2 = a_1 x_1 + 1 = a_1 [a_0 x_0 + 1] + 1 ,$$

$$= a_1 a_0 x_0 + a_1 + 1 = -x_0 + a_1 + 1 ,$$

$$x_3 = a_0 x_2 + 1 = a_0 [-x_0 + a_1 + 1] + 1$$

$$= -a_0 x_0 + a_0 a_1 + a_0 + 1 = -a_0 x_0 + a_0$$

$$x_4 = a_1 x_3 + 1 = a_1 [-a_0 x_0 + a_0] + 1 = x_0 .$$

Period-4 solutions are possible if and only if $a_0 a_1 = -1$ with the following period-4 pattern:

$$x_0 , \ a_0 x_0 + 1 , \ -x_0 + a_1 + 1 , \ -a_0 x_0 + a_0, \ \ldots .$$

Next we will transition to the periodicity nature of solutions of:

$$x_{n+1} = -x_n + a_n \quad, \quad n = 0, 1, \ldots , \tag{2.8}$$

where $\{a_n\}_{n=0}^{\infty}$ is a period-k sequence, $(k \geq 2)$. We will discover when $\{a_n\}_{n=0}^{\infty}$ is of odd order then Equation (2.8) will have unique periodic solutions and when $\{a_n\}_{n=0}^{\infty}$ is of even order then every solution of Equation (2.8) is periodic.

Example 14. Suppose that $\{a_n\}_{n=0}^{\infty}$ is a period-2 sequence. Show that:

$$x_{n+1} = -x_n + a_n \ , \quad n = 0, 1, \ldots \ ,$$

has no period-2 solutions and explain why.

Solution: Suppose that $x_2 = x_0$. Then:

$$x_0 \ ,$$

$$x_1 = -x_0 + a_0 \ ,$$

$$x_2 = -[x_1] + a_1 \ = \ x_0 - a_0 + a_1 \ = \ x_0 \ .$$

Therefore $x_2 = x_0$ if and only if $a_0 = a_1$. This is a contradiction as we assumed that $\{a_n\}_{n=0}^{\infty}$ is a period-2 sequence.

Example 15. Suppose that $\{a_n\}_{n=0}^{\infty}$ is a period-3 sequence. Determine the necessary and sufficient conditions for the existence of period-3 cycles of:

$$x_{n+1} = -x_n + a_n \ , \quad n = 0, 1, \ldots \ .$$

Solution: Observe:

$$x_0 \ ,$$

$$x_1 = -x_0 + a_0 \ ,$$

$$x_2 = -[x_1] + a_1 \ = \ -[-x_0 + a_0] + a_1 \ = \ x_0 + a_1 - a_0 \ ,$$

$$x_3 = -[x_2] + a_2 \ = \ -[x_0 + a_1 - a_0] + a_2 \ = \ -x_0 + a_2 + a_0 - a_1 \ = \ x_0 \ .$$

Therefore $x_0 = \frac{a_0 - a_1 + a_2}{2}$ and acquire the following **unique** period-3 pattern:

$$\frac{a_0 - a_1 + a_2}{2} \ , \ \frac{a_0 + a_1 - a_2}{2} \ , \ \frac{-a_0 + a_1 + a_2}{2} \ \ldots \ .$$

Note that from neighbor to neighbor the indices do not change but the negative sign shifts from neighbor to neighbor in the numerator only.

Example 16. Suppose that $\{a_n\}_{n=0}^{\infty}$ is a period-4 sequence. Determine the necessary and sufficient conditions for the existence of period-4 cycles of:

$$x_{n+1} = -x_n + a_n \ , \quad n = 0, 1, \ldots \ .$$

Solution: By iteration:

$$x_0 \ ,$$

$$x_1 = -x_0 + a_0 \ ,$$

$$x_2 = -[x_1] + a_1 \ = \ -[-x_0 + a_0] + a_1 \ = \ x_0 + a_1 - a_0 \ ,$$

$$x_3 = -[x_2] + a_2 \ = \ -[x_0 + a_1 - a_0] + a_2 \ = \ -x_0 + a_2 + a_0 - a_1$$

$$x_4 = -[x_3] + a_3 \ = \ -[-x_0 + a_2 + a_0 - a_1] + a_3 \ ,$$

$$= \ x_0 - [a_2 + a_0] + [a_1 + a_3] \ = \ x_0 \ .$$

Period-4 solutions exist if and only if $a_1 + a_3 = a_0 + a_2$ with the following period-4 pattern:

$$x_0 \ , \ -x_0 + a_0 \ , \ x_0 + a_1 - a_0 \ , \ -x_0 + a_2 + a_0 - a_1, \ \ldots \ .$$

From Example 15 and Example 16 we can observe quite a contrast when $\{a_n\}_{n=0}^{\infty}$ is an even ordered periodic sequence and when $\{a_n\}_{n=0}^{\infty}$ is an odd ordered periodic sequence. First of all, in Example 15 we have a unique periodic cycle when $\{a_n\}_{n=0}^{\infty}$ is an odd ordered periodic sequence. Second, in Example 16 every solution is periodic when $\{a_n\}_{n=0}^{\infty}$ is an even-ordered periodic sequence. The following two theorems generalize the conclusions.

Theorem 4. *Suppose that* $\{a_n\}_{n=0}^{\infty}$ *is a period-2k sequence, $(k \geq 2)$. Then every solution of:*

$$x_{n+1} \ = \ -x_n + a_n \ , \ \ n = 0, 1, \ldots \ ,$$

is periodic with period-2k if and only if:

$$\sum_{i=1}^{k} a_{2i-1} \ = \ \sum_{i=1}^{k} a_{2i-2} \ .$$

Proof: *Similar to Example 15, we get:*

$$x_0 \ ,$$

$$x_1 = -x_0 + a_0 \ ,$$

$$x_2 = -[x_1] + a_1 = x_0 + a_1 - a_0 \ ,$$

$$x_3 = -[x_2] + a_2 = -x_0 + a_2 + a_0 - a_1 \ ,$$

$$x_4 = -[x_3] + a_3 = x_0 - [a_2 + a_0] + [a_1 + a_3] \ ,$$

$$\vdots$$

$$x_{2k} = x_0 - \left[\sum_{i=1}^{k} a_{2i-2} \right] + \left[\sum_{i=1}^{k} a_{2i-1} \right] \ .$$

Theorem 5. *Suppose that $\{a_n\}_{n=0}^{\infty}$ is a period-$(2k+1)$ sequence, $(k \in \mathbb{N})$. Then:*

$$x_{n+1} = -x_n + a_n \quad , \quad n = 0, 1, \ldots ,$$

has a unique periodic solution with period-$(2k+1)$ where:

$$x_0 = \frac{\sum_{i=1}^{2k+1} (-1)^{i+1} a_{i-1}}{2} .$$

The proof of Theorem 5 will be left as an exercise at the end of the chapter. Now we will shift our analysis of periodic traits of:

$$x_{n+1} = a_n x_n + b_n \quad , \quad n = 0, 1, \ldots , \tag{2.9}$$

where $\{a_n\}_{n=0}^{\infty}$ and $\{b_n\}_{n=0}^{\infty}$ are periodic sequences of either the same period or with different periods. The next two examples will show unique periodic cycles of Equation (2.9) and their patterns.

Example 17. Suppose that $\{a_n\}_{n=0}^{\infty}$ and $\{b_n\}_{n=0}^{\infty}$ are period-2 sequences. Determine the pattern of the unique period-2 cycle of:

$$x_{n+1} = a_n x_n + b_n \quad , \quad n = 0, 1, \ldots .$$

Solution: First set $x_2 = x_0$ and by iteration we get:

$$x_0 ,$$

$$x_1 = a_0 x_0 + b_0 ,$$

$$x_2 = a_1 [x_1] + b_1 = a_1 [a_0 x_0 + b_0] + b_1 = a_0 a_1 x_0 + a_1 b_0 + b_1 = x_0 .$$

We solve for x_0 and acquire the following unique period-2 pattern:

$$\frac{a_1 b_0 + b_1}{1 - a_0 a_1} , \quad \frac{a_0 b_1 + b_0}{1 - a_0 a_1} , \quad \frac{a_1 b_0 + b_1}{1 - a_0 a_1} , \quad \frac{a_0 b_1 + b_0}{1 - a_0 a_1} , \quad \ldots ,$$

provided that $a_0 a_1 \neq 1$. First of all, we have a unique period-2 cycle. Second, the indices of the denominator are the same from term to term and the indices of the numerator shift by an index of one from term to term.

Example 18. Suppose that $\{a_n\}_{n=0}^{\infty}$ and $\{b_n\}_{n=0}^{\infty}$ are period-3 sequences. Determine the pattern of the unique period-3 cycle of:

$$x_{n+1} = a_n x_n + b_n \quad , \quad n = 0, 1, \ldots .$$

Solution: Set $x_3 = x_0$ and by iteration we get:

$$x_0 \, ,$$

$$x_1 = a_0 x_0 + b_0 \, ,$$

$$x_2 = a_1 [x_1] + b_1 = a_1 [a_0 x_0 + b_0] + b_1 = a_0 a_1 x_0 + a_1 b_0 + b_1 \, ,$$

$$x_3 = a_2 [x_2] + b_2 = a_2 [a_0 a_1 x_0 + a_1 b_0 + b_1] + b_2 \, ,$$

$$= a_0 a_1 a_2 x_0 + a_1 a_2 b_0 + a_2 b_1 + b_2 = x_0 \, .$$

We solve for x_0 and get

$$x_0 \;=\; \frac{a_1 a_2 b_0 + a_2 b_1 + b_2}{1 - a_0 a_1 a_2} \;=\; \frac{a_2 [a_1 b_0 + b_1] + b_2}{1 - a_0 a_1 a_2} \, .$$

Note that $a_0 a_1 a_2 \neq 1$ and we procure the following period-3 pattern:

$$\frac{a_2 [a_1 b_0 + b_1] + b_2}{1 - a_0 a_1 a_2} \, , \; \frac{a_0 [a_2 b_1 + b_2] + b_0}{1 - a_0 a_1 a_2} \, , \; \frac{a_1 [a_0 b_2 + b_0] + b_1}{1 - a_0 a_1 a_2} \, , \; \dots \, .$$

Similar to the previous example, we have a unique period-3 cycle. The coefficients of the denominator are the same from term to term and the coefficients of the numerator shift by an index of one.

From Examples 17 and 18 we can extend the result when $\{a_n\}_{n=0}^{\infty}$ and $\{b_n\}_{n=0}^{\infty}$ are both periodic-k sequences, $(k \geq 2)$. The following theorem outlines the conclusions.

Theorem 6. *Suppose that $\{a_n\}_{n=0}^{\infty}$ and $\{b_n\}_{n=0}^{\infty}$ are periodic sequences with period-k, $(k \geq 2)$. Then:*

$$x_{n+1} \;=\; a_n x_n + b_n \, , \quad n = 0, 1, \dots \, ,$$

has a unique period-k cycle if

$$\prod_{i=0}^{k-1} a_i \neq 1 \, ,$$

and

$$x_0 \;=\; \frac{a_{k-1} [\dots [a_3 [a_2 [a_1 b_0 + b_1] + b_2] + b_3] \dots] + b_{k-1}}{1 - \left[\prod_{i=0}^{k-1} a_i \right]} \, .$$

The proof of Theorem 6 will be left as an exercise at the end of the chapter. Furthermore, the **Open Problem** below will address what happens to the periodic character of solutions when $\{a_n\}_{n=0}^{\infty}$ and $\{b_n\}_{n=0}^{\infty}$ are periodic sequences with different periods?

Open Problem 7. *Suppose that $\{a_n\}_{n=0}^{\infty}$ is periodic with period-k, $(k \geq 2)$, and $\{b_n\}_{n=0}^{\infty}$ is periodic with period-l, $(l \geq 2)$ and $k \neq l$. Discuss the periodic character of:*

$$x_{n+1} \;=\; a_n x_n + b_n \, , \quad n = 0, 1, \dots \, ,$$

To address this question, we will need to investigate different cases when k is a multiple of l and when k and l are relatively prime. Now we will proceed with our study of first order linear nonautonomous Δ.E. in the form:

$$x_{n+1} = ax_n + b_n \quad , \quad n = 0, 1, \dots , \tag{2.10}$$

where $x_0 \in \mathfrak{R}$, $a \neq 1$, and $\{b_n\}_{n=0}^{\infty}$ is a sequence of real numbers. Let $\{x_n\}_{n=0}^{\infty}$ be a solution of Equation (2.10). By iteration we acquire:

$$x_1 = ax_0 + b_0 ,$$

$$x_2 = ax_1 + b_1 = a^2 x_0 + ab_0 + b_1 ,$$

$$x_3 = ax_2 + b_2 = a^3 x_0 + a^2 b_0 + ab_1 + b_2 ,$$

$$x_4 = ax_3 + b_3 = a^4 x_0 + a^3 b_0 + a^2 b_1 + ab_2 + b_3 ,$$

$$x_5 = ax_4 + b_4 = a^5 x_0 + a^4 b_0 + a^3 b_1 + a^2 b_2 + ab_3 + b_4 .$$

$$\vdots$$

Hence for all $n \in \mathbb{N}$,

$$x_n = a^n x_0 + \left(a^{n-1} b_0 + a^{n-2} b_1 + a^{n-3} b_2 + \dots + ab_{n-2} + b_{n-1} \right)$$

$$= a^n x_0 + \sum_{k=0}^{n-1} a^{(n-1-k)} b_k .$$

In the next two examples, we will work out the details solving the nonautonomous first order linear difference equations.

Example 19. Solve the following Initial Value Problem:

$$\begin{cases} x_{n+1} - x_n = n \quad , \quad n = 0, 1, \dots, \\ \\ x_0 = 4 . \end{cases}$$

Solution: From Equation (2.10), we obtain:

$$x_n = x_0 + \sum_{k=0}^{n-1} k = 4 + \left[\frac{n(n+1)}{2} \right] .$$

Example 20. Solve the following Initial Value Problem:

$$\begin{cases} x_{n+1} - x_n = 2^n \quad , \quad n = 0, 1, \dots, \\ \\ x_0 = 1 . \end{cases}$$

Solution: Using Equation (2.10), we procure:

$$x_n = x_0 + \sum_{k=0}^{n-1} 2^k = x_0 + \frac{1 - 2^n}{1 - 2} = x_0 + [2^n - 1].$$

2.5 Applications of First Order Linear Difference Equations in Biology

We will commence with applications of first order linear homogeneous difference equations by examining the population of a specie that either increases or decreases at a constant rate of r % per year. We define:

$$x_n = \text{the population at year n,}$$

$$x_{n+1} = \text{the population at year n+1,}$$

$$P_0 = \text{the initial population.}$$

Then we see that when the population **increases** at a constant rate of r % per year, we acquire the following initial value problem:

$$\begin{cases} x_{n+1} = x_n + \left(\frac{r}{100}\right) x_n , & n = 0, 1, \ldots \\ \\ x_0 = P_0 . \end{cases}$$

Similarly, when the population **decreases** at a constant rate of r % per year, we get the following initial value problem:

$$\begin{cases} x_{n+1} = x_n - \left(\frac{r}{100}\right) x_n , & n = 0, 1, \ldots \\ \\ x_0 = P_0 . \end{cases}$$

2.6 Applications of First Order Linear Difference Equations in Finance

Now we shift our aim in applications of first order linear difference equations in compound interest rates; in particular, investing an initial amount of money compounded number of times per year and paying off loans. Consider investing Q_0 dollars at a rate of r % per year, compounded k times per year. Then we get the following initial value problem:

$$\begin{cases} x_{n+1} = x_n + \left(\frac{r}{k}\right)x_n = \left(1+\frac{r}{k}\right)x_n \quad , \quad n=0,1,\ldots \\ \\ x_0 = P_0 \, . \end{cases}$$

The following theorem describes the periodic payment required to pay off a loan with a constant interest rate.

Theorem 8. *Suppose that L dollars of loan is taken out. In addition, the loan pay off is divided into N periods at a constant rate of i % per year. Then, the periodic payment P is*

$$P = \frac{iL}{1-(1+i)^{-N}} \, .$$

Proof: *Let x_n denote the amount owed immediately after the nth payment. Then we get the following initial value problem:*

$$\begin{cases} x_{n+1} = x_n + ix_n - P \quad , \quad n=0,1,\ldots,N, \\ \\ x_0 = L \, , \\ \\ x_N = 0 \, . \end{cases}$$

Therefore via (2.5), the solution of the above initial value problem is:

$$x_n = L(1+i)^n + \left(\frac{P}{i}\right)[1-(1+i)^n] \quad , \quad n=0,1,\ldots,N.$$

By setting $x_N = 0$, we acquire:

$$L(1+i)^N + \left(\frac{P}{i}\right)[1-(1+i)^N] = 0 \, .$$

The result follows.

2.7 Exercises

In problems 1–5, show that the solution **satisfies** the given Δ.E.:

1. $x_n = \left(\frac{2}{3}\right)^{n+2}$ is a solution of $3x_{n+1} - 2x_n = 0$, $n=0,1,\ldots$.

2. $x_n = 2+6n$ is a solution of $x_{n+1} = x_n + 6$, $n=0,1,\ldots$.

3. $x_n = \frac{3}{2}$ is a solution of $2x_{n+1} - 4x_n = -3$, $n=0,1,\ldots$.

4. $x_n = (-3)^{n+1} + 1$ is a solution of $x_{n+1} + 3x_n = 4$, $n=0,1,\ldots$

5. $x_n = \left(\frac{3}{4}\right)^{n+2} + 4$ is a solution of $4x_{n+1} - 3x_n = 4$, $n=0,1,\ldots$

In problems 6–18, solve the given **Initial Value Problem** and check your answer.

6.
$$\begin{cases} 3x_{n+1} - 2x_n = 0 \ , \quad n = 0,1,\ldots. \\ \\ x_0 = \frac{4}{9} \ . \end{cases}$$

7.
$$\begin{cases} 4x_{n+1} + 5x_n = 0 \ , \quad n = 0,1,\ldots. \\ \\ x_0 = \frac{6}{5} \ . \end{cases}$$

8.
$$\begin{cases} 2x_{n+1} + 2x_n = 5 \ , \quad n = 0,1,\ldots. \\ \\ x_0 = \frac{3}{4} \ . \end{cases}$$

9.
$$\begin{cases} x_{n+1} - x_n = -3 \ , \quad n = 0,1,\ldots. \\ \\ x_0 = -1 \ . \end{cases}$$

10.
$$\begin{cases} 4x_{n+1} - 4x_n = 1 \ , \quad n = 0,1,\ldots. \\ \\ x_0 = \frac{3}{4} \ . \end{cases}$$

11.
$$\begin{cases} x_{n+1} + x_n = 5 \ , \quad n = 0,1,\ldots. \\ \\ x_0 = -2 \ . \end{cases}$$

12.
$$\begin{cases} x_{n+1} - x_n = \left(\frac{2}{3}\right)^{n+1} \ , \quad n = 0,1,\ldots. \\ \\ x_0 = \frac{3}{2} \ . \end{cases}$$

13.
$$\begin{cases} x_{n+1} + x_n = 2^n \ , \quad n = 0,1,\ldots. \\ \\ x_0 = 4 \ . \end{cases}$$

14.
$$\begin{cases} x_{n+1} = x_n + (n+1)^2 \ , \quad n = 0,1,\ldots. \\ \\ x_0 = 2 \ . \end{cases}$$

15.
$$\begin{cases} x_{n+1} = x_n + 4n \ , \quad n = 0,1,\ldots. \\ \\ x_0 = 0 \ . \end{cases}$$

16.
$$\begin{cases} x_{n+1} = x_n + n + n^2 \ , & n = 0, 1, \ldots . \\ \\ x_0 = 3 . \end{cases}$$

17.
$$\begin{cases} x_{n+1} = x_n + 4n + 2^n \ , & n = 0, 1, \ldots . \\ \\ x_0 = 2 . \end{cases}$$

18.
$$\begin{cases} x_{n+1} + x_n = n \ , & n = 0, 1, \ldots . \\ \\ x_0 = 1 . \end{cases}$$

In problems 19–28, determine the general solution to each Δ.E. In addition, determine if the limit exists of each solution. If so, determine if the limit is unique. If not, explain why.

19. $4x_{n+1} + 2x_n = 1$, $n = 0, 1, \ldots$.

20. $3x_{n+1} - 6x_n = 4$, $n = 0, 1, \ldots$.

21. $6x_{n+1} - 4x_n = 1$, $n = 0, 1, \ldots$.

22. $3x_{n+1} + 3x_n = 2$, $n = 0, 1, \ldots$.

23. $2x_{n+1} + 5x_n = 0$, $n = 0, 1, \ldots$.

24. $2x_{n+1} - 2x_n = 3$, $n = 0, 1, \ldots$.

25. $3x_{n+1} + 3x_n = -2$, $n = 0, 1, \ldots$.

26. $x_{n+1} - x_n = \frac{1}{(n+1)!}$, $n = 0, 1, \ldots$.

27. $x_{n+1} + x_n = \frac{(-1)^n}{n+1}$, $n = 0, 1, \ldots$.

28. $x_{n+1} - x_n = \frac{3^{n+1}}{4^n}$, $n = 0, 1, \ldots$.

29. Suppose that 600 dollars is invested at a constant rate of 6 percent per year compounded quarterly. Write the initial value problem that describes the dynamics and determine how long it will take to reach 10,000 dollars.

30. Suppose that 250 dollars is invested at a constant rate of 2 percent per year compounded monthly. Write the initial value problem that describes the dynamics and determine how long it will take to reach 8,000 dollars.

31. Suppose that 20,000 dollar loan is taken out of a bank at an interest rate i. Suppose that the loan must be paid off in 10 years. Write the initial value problem that describes the dynamics and then use it to determine i.

32. Suppose that 16,000 dollar loan is taken out of a bank at an interest rate of 4 percent. Write the initial value problem that describes the dynamics and then use it to determine how long it will take to pay off the loan.

In problems 33–53, determine the necessary and sufficient conditions for the existence of periodic solutions and the pattern of periodic solutions.

33. Existence and Pattern of Period-8 Solutions of the Δ.E.:

$$x_{n+1} = (-1)^n x_n + 4 \quad , \quad n = 0, 1, \dots .$$

34. Existence and Pattern of Period-8 Solutions of the Δ.E.:

$$x_{n+1} = -a_n x_n + 1 \quad , \quad n = 0, 1, \dots ,$$

where $\{a_n\}_{n=0}^{\infty}$ is a period-4 sequence.

35. Existence and Pattern of Period-10 Solutions of the Δ.E.:

$$x_{n+1} = a_n x_n - 1 \quad , \quad n = 0, 1, \dots ,$$

where $\{a_n\}_{n=0}^{\infty}$ is a period-5 sequence.

36. Existence and Pattern of Period-12 Solutions of the Δ.E.:

$$x_{n+1} = x_n + a_n \quad , \quad n = 0, 1, \dots ,$$

where $\{a_n\}_{n=0}^{\infty}$ is a period-12 sequence.

37. Existence and Pattern of Period-k Solutions of the Δ.E.:

$$x_{n+1} = x_n + a_n \quad , \quad n = 0, 1, \dots ,$$

where $\{a_n\}_{n=0}^{\infty}$ is a period-k sequence, $(k \geq 2)$.

38. Existence and Pattern of Period-5 Solutions of the Δ.E.:

$$x_{n+1} = -x_n + a_n \quad , \quad n = 0, 1, \dots ,$$

where $\{a_n\}_{n=0}^{\infty}$ is a period-5 sequence.

39. Existence and Pattern of Period-6 Solutions of the Δ.E.:

$$x_{n+1} = -x_n + a_n \quad , \quad n = 0, 1, \dots ,$$

where $\{a_n\}_{n=0}^{\infty}$ is a period-6 sequence.

40. Existence and Pattern of Period-2k Solutions of the Δ.E.:

$$x_{n+1} = -x_n + a_n \quad , \quad n = 0, 1, \ldots ,$$

where $\{a_n\}_{n=0}^{\infty}$ is a period-2k sequence, $(k \in \mathbb{N})$.

41. Existence and Pattern of Period-$(2k+1)$ Solutions of the Δ.E.:

$$x_{n+1} = -x_n + a_n \quad , \quad n = 0, 1, \ldots ,$$

where $\{a_n\}_{n=0}^{\infty}$ is a period-$(2k+1)$ sequence, $(k \in \mathbb{N})$.

42. Existence and Pattern of Period-4 Solutions of the Δ.E.:

$$x_{n+1} = a_n x_n + (-1)^{n+1} \quad , \quad n = 0, 1, \ldots ,$$

where $\{a_n\}_{n=0}^{\infty}$ is a period-2 sequence.

43. Existence and Pattern of Period-8 Solutions of the Δ.E.:

$$x_{n+1} = a_n x_n + (-1)^{n+1} \quad , \quad n = 0, 1, \ldots ,$$

where $\{a_n\}_{n=0}^{\infty}$ is a period-4 sequence.

44. Existence and Pattern of Period-2k Solutions of the Δ.E.:

$$x_{n+1} = -a_n x_n \quad , \quad n = 0, 1, \ldots ,$$

where $\{a_n\}_{n=0}^{\infty}$ is a period-k sequence, $(k \geq 2)$.

45. Existence and Pattern of Period-4 Solutions of the Δ.E.:

$$x_{n+1} = a_n x_n + b_n \quad , \quad n = 0, 1, \ldots ,$$

where $\{a_n\}_{n=0}^{\infty}$ and $\{b_n\}_{n=0}^{\infty}$ are period-4 sequences.

46. Existence and Pattern of Period-5 Solutions of the Δ.E.:

$$x_{n+1} = a_n x_n + b_n \quad , \quad n = 0, 1, \ldots ,$$

where $\{a_n\}_{n=0}^{\infty}$ and $\{b_n\}_{n=0}^{\infty}$ are period-5 sequences.

47. Existence and Pattern of Period-k Solutions of the Δ.E.:

$$x_{n+1} = a_n x_n + b_n \quad , \quad n = 0, 1, \ldots ,$$

where $\{a_n\}_{n=0}^{\infty}$ and $\{b_n\}_{n=0}^{\infty}$ are period-k sequences, $(k \geq 2)$.

48. Existence and Pattern of Periodic Solutions of the Δ.E.:

$$x_{n+1} = a_n x_n + b_n \quad , \quad n = 0, 1, \dots ,$$

where $\{a_n\}_{n=0}^{\infty}$ is period-2 sequence and $\{b_n\}_{n=0}^{\infty}$ is period-4 sequence.

49. Existence and Pattern of Periodic Solutions of the Δ.E.:

$$x_{n+1} = a_n x_n + b_n \quad , \quad n = 0, 1, \dots ,$$

where $\{a_n\}_{n=0}^{\infty}$ is period-4 sequence and $\{b_n\}_{n=0}^{\infty}$ is period-2 sequence.

50. Existence and Pattern of Periodic Solutions of the Δ.E.:

$$x_{n+1} = a_n x_n + b_n \quad , \quad n = 0, 1, \dots ,$$

where $\{a_n\}_{n=0}^{\infty}$ is period-2k sequence ($k \in \mathbb{N}$), $\{b_n\}_{n=0}^{\infty}$ is period-2l sequence ($l \in \mathbb{N}$), and $k \neq l$.

51. Existence and Pattern of Periodic Solutions of the Δ.E.:

$$x_{n+1} = a_n x_n + b_n \quad , \quad n = 0, 1, \dots ,$$

where $\{a_n\}_{n=0}^{\infty}$ is period-2 sequence and $\{b_n\}_{n=0}^{\infty}$ is period-3 sequence.

52. Existence and Pattern of Periodic Solutions of the Δ.E.:

$$x_{n+1} = a_n x_n + b_n \quad , \quad n = 0, 1, \dots ,$$

where $\{a_n\}_{n=0}^{\infty}$ is period-3 sequence and $\{b_n\}_{n=0}^{\infty}$ is period-2 sequence.

53. Existence and Pattern of Periodic Solutions of the Δ.E.:

$$x_{n+1} = a_n x_n + b_n \quad , \quad n = 0, 1, \dots ,$$

where $\{a_n\}_{n=0}^{\infty}$ is period-k sequence (for $k \geq 2$), $\{b_n\}_{n=0}^{\infty}$ is period-l sequence (for $l \geq 2$), and $k \neq l$.

Problems 54–62 are open-ended questions. Determine if periodic solutions exist. If so, then determine the pattern of the period. If not, then explain why. Determine either analytically or from computer observations.

54. Periodic Solutions of the Δ.E.:

$$x_{n+1} = (-1)^n x_n + a_n \quad , \quad n = 0, 1, \dots ,$$

where $\{a_n\}_{n=0}^{\infty}$ is period-2k sequence, ($k \in \mathbb{N}$).

55. Periodic Solutions of the Δ.E.:

$$x_{n+1} = (-1)^n x_n + a_n \quad , \quad n = 0, 1, \dots ,$$

where $\{a_n\}_{n=0}^{\infty}$ is period-$(2k+1)$ sequence, $(k \in \mathbb{N})$.

56. Periodic Solutions of the Δ.E.:

$$x_{n+1} = (-1)^n a_n x_n + b_n \quad , \quad n = 0, 1, \dots ,$$

where $\{a_n\}_{n=0}^{\infty}$ and $\{b_n\}_{n=0}^{\infty}$ are period-k sequences (for $k \geq 2$).

57. Periodic Solutions of the Δ.E.:

$$x_{n+1} = (-1)^n a_n x_n + b_n \quad , \quad n = 0, 1, \dots ,$$

where $\{a_n\}_{n=0}^{\infty}$ is a period-k sequence (for $k \geq 2$), $\{b_n\}_{n=0}^{\infty}$ is a period-l sequence (for $l \geq 2$) and $k \neq l$.

58. Periodic Solutions of the Δ.E.:

$$x_{n+1} = a_n x_n + b_n + c_n \quad , \quad n = 0, 1, \dots ,$$

where $\{a_n\}_{n=0}^{\infty}$, $\{b_n\}_{n=0}^{\infty}$ and $\{c_n\}_{n=0}^{\infty}$ are period-2 sequences.

59. Periodic Solutions of the Δ.E.:

$$x_{n+1} = a_n x_n + b_n + c_n \quad , \quad n = 0, 1, \dots ,$$

where $\{a_n\}_{n=0}^{\infty}$, $\{b_n\}_{n=0}^{\infty}$ and $\{c_n\}_{n=0}^{\infty}$ are period-3 sequences.

60. Periodic Solutions of the Δ.E.:

$$x_{n+1} = a_n x_n + b_n + c_n \quad , \quad n = 0, 1, \dots ,$$

where $\{a_n\}_{n=0}^{\infty}$, $\{b_n\}_{n=0}^{\infty}$ and $\{c_n\}_{n=0}^{\infty}$ are period-k sequences (for $k \geq 2$).

61. Periodic Solutions of the Δ.E.:

$$x_{n+1} = a_n x_n + b_n + c_n \quad , \quad n = 0, 1, \dots ,$$

where $\{a_n\}_{n=0}^{\infty}$ and $\{b_n\}_{n=0}^{\infty}$ are period-2 sequences and $\{c_n\}_{n=0}^{\infty}$ is a period-3 sequence.

62. Periodic Solutions of the Δ.E.:

$$x_{n+1} = a_n x_n + b_n + c_n \quad , \quad n = 0, 1, \dots ,$$

where $\{a_n\}_{n=0}^{\infty}$ and $\{b_n\}_{n=0}^{\infty}$ are period-3 sequences and $\{c_n\}_{n=0}^{\infty}$ is a period-2 sequence.

Problems 63–71 are open-ended questions. Using a computer, determine the periodic character of solutions to the following linear difference equations with **modulo arithmetic** and the initial condition $x_0 = 1$.

63. $x_{n+1} = 2x_n \ (Mod \ 3)$, $n = 0, 1, \ldots$.

64. $x_{n+1} = 2x_n \ (Mod \ 4)$, $n = 0, 1, \ldots$.

65. $x_{n+1} = 2x_n \ (Mod \ 5)$, $n = 0, 1, \ldots$.

66. $x_{n+1} = 2x_n \ (Mod \ 6)$, $n = 0, 1, \ldots$.

65. $x_{n+1} = 2x_n \ (Mod \ 7)$, $n = 0, 1, \ldots$.

67. $x_{n+1} = 3x_n \ (Mod \ 2)$, $n = 0, 1, \ldots$.

68. $x_{n+1} = 3x_n \ (Mod \ 4)$, $n = 0, 1, \ldots$.

69. $x_{n+1} = 3x_n \ (Mod \ 5)$, $n = 0, 1, \ldots$.

70. $x_{n+1} = 3x_n \ (Mod \ 6)$, $n = 0, 1, \ldots$.

71. $x_{n+1} = 3x_n \ (Mod \ 7)$, $n = 0, 1, \ldots$.

Chapter 3
Periodic Character of Solutions of First Order Nonlinear Difference Equations

3.1 First Order Nonlinear Difference Equations

This chapter's aims are to determine the necessary and sufficient criteria for the existence of periodic solutions, patterns of periodic solutions, and the existence of eventually periodic solutions of the first order nonlinear difference equations. We emerge with three examples of first order nonlinear difference equations that exhibit periodic solutions and eventually periodic solutions:

(i) (Special Case of the **Riccati Δ.E.**)

$$x_{n+1} = \frac{x_n}{x_n - 1} \quad , \quad n = 0, 1, \dots .$$

(ii) (Special Case of the **Logistic Δ.E.**)

$$x_{n+1} = 4x_n(1 - x_n) \quad , \quad n = 0, 1, \dots .$$

(iii) (Special Case of the piece-wise difference equations **Tent-Map**)

$$x_{n+1} = \begin{cases} 2x_n & \text{if } x_n < \frac{1}{2} , \\[2mm] 2(1 - x_n) & \text{if } x_n \geq \frac{1}{2} . \end{cases} \quad n = 0, 1, \dots ,$$

Recall from Chapter 1, the sequence $\{x_n\}_{n=0}^{\infty}$ is periodic with **minimal period-p≥ 2**, provided that

$$x_{n+p} = x_n \quad \text{for all} \quad n = 0, 1, \dots .$$

We will commence our study of the existence and patterns of periodic solutions of the **Riccati Difference Equation**.

© Springer Nature Switzerland AG 2018
M. A. Radin, *Periodic Character and Patterns of Recursive Sequences*,
https://doi.org/10.1007/978-3-030-01780-4_3

3.2 The Riccati Difference Equation

First we will start with the **Riccati Δ.E.**:

$$x_{n+1} = \frac{ax_n + b}{cx_n + d} \quad, \quad n = 0, 1, \ldots ,$$

where the parameters $a, b, c, d \geq 0$ and $x_0 \geq 0$. Riccati Δ.E. has many applications in gravitational sciences and biological sciences. Observe that by setting:

$$\bar{x} = \frac{a\bar{x} + b}{c\bar{x} + d} ,$$

we get the following quadratic equation:

$$c\bar{x}^2 + (d - a)\bar{x} - b = 0 ,$$

with two solutions:

$$\bar{x}_1 = \frac{(a-d) + \sqrt{(a-d)^2 + 4bc}}{2c} \quad \text{and} \quad \bar{x}_2 = \frac{(a-d) - \sqrt{(a-d)^2 + 4bc}}{2c} .$$

Now we examine the existence of period-2 cycles by setting $x_2 = x_0$ and we get:

$$
\begin{aligned}
x_0 \, , & \\
x_1 &= \frac{ax_0 + b}{cx_0 + d} , \\
x_2 &= \frac{ax_1 + b}{cx_1 + d} = \frac{a\left[\frac{ax_0 + b}{cx_0 + d}\right] + b}{c\left[\frac{ax_0 + b}{cx_0 + d}\right] + d} = x_0 ,
\end{aligned}
$$

which gives us:

$$\frac{a^2 x_0 + ab + bcx_0 + bd}{acx_0 + bc + dcx_0 + d^2} = x_0 ,$$

and reduces to:

$$c(a+d)x_0^2 + (d^2 - a^2)x_0 - b(a+d) = 0 ,$$

with two solutions:

$$x_0 = \frac{(a-d) + \sqrt{(a-d)^2 + 4bc}}{2c} \quad \text{and} \quad x_0 = \frac{(a-d) - \sqrt{(a-d)^2 + 4bc}}{2c} .$$

The two solutions above are the equilibrium points $\bar{x}_1$ and $\bar{x}_2$. We proceed with studying periodic traits of two special cases of the Riccati Δ.E. where every non-trivial solution (nonequilibrium solution) is periodic with period-2.

$$x_{n+1} = \frac{1}{x_n} \quad, \quad n = 0, 1, \ldots , \tag{3.1}$$

and

$$x_{n+1} = \frac{x_n}{x_n - 1} \quad , \quad n = 0, 1, \ldots \, , \tag{3.2}$$

We will emerge with the periodic character of solutions of Equation (3.1) and assume that $x_0 \neq 0$.

Example 1. Show that every nontrivial solution of:

$$x_{n+1} = \frac{1}{x_n} \quad , \quad n = 0, 1, \ldots \, ,$$

is periodic with period-2 and determine the pattern of period-2 cycles; ($x_0 \neq 0$).

Solution: By iteration we acquire:

$$x_0 \, ,$$

$$x_1 = \frac{1}{x_0} \, ,$$

$$x_2 = \frac{1}{x_1} = \frac{1}{\left[\frac{1}{x_0}\right]} = x_0 \, ,$$

$$x_3 = \frac{1}{x_2} = \frac{1}{x_0} = x_1 \, .$$

Hence we obtain the following period-2 pattern:

$$x_0 \, , \, \frac{1}{x_0} \, , \, x_0 \, , \, \frac{1}{x_0} \, , \, \ldots \, .$$

Now notice that the product of two neighboring terms:

$$x_0 \cdot \frac{1}{x_0} = 1 \, .$$

The periodic cycles are on the hyperbolic curve $y = \frac{1}{x}$ (Figure 3.1):

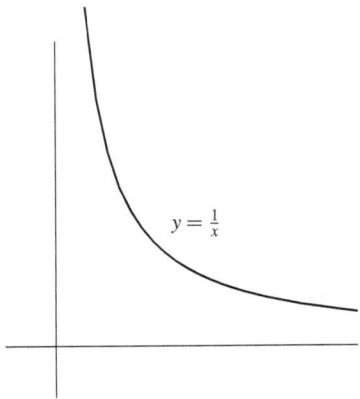

$$y = \frac{1}{x}$$

Fig. 3.1 Graph of $y = \frac{1}{x}$.

Now we proceed with the periodic nature of solutions of Equation (3.2) and suppose that $x_0 \neq 1$.

Example 2. Show that every nontrivial solution of:

$$x_{n+1} = \frac{x_n}{x_n - 1} \quad , \quad n = 0, 1, \dots ,$$

is periodic with period-2 and determine the pattern of period-2 cycles; $(x_0 \neq 1)$
Solution: Notice:

$$x_0 ,$$

$$x_1 = \frac{x_0}{x_0 - 1} ,$$

$$x_2 = \frac{x_1}{x_1 - 1} = \frac{\left[\frac{x_0}{x_0-1}\right]}{\left[\frac{x_0}{x_0-1}\right] - 1} = \frac{x_0}{x_0 - (x_0 - 1)} = x_0 ,$$

$$x_3 = \frac{x_2}{x_2 - 1} = \frac{x_0}{x_0 - 1} = x_1 .$$

We then procure the following period-2 pattern:

$$x_0 , \frac{x_0}{x_0 - 1} , x_0 , \frac{x_0}{x_0 - 1} , \dots .$$

Notice that the product and the sum of two neighboring terms are always equal:

$$x_0 \cdot \frac{x_0}{x_0 - 1} = \frac{(x_0)^2}{x_0 - 1} = x_0 + \frac{x_0}{x_0 - 1} .$$

The periodic solutions are on the hyperbolic curve $y = \frac{x}{x-1}$ (Figure 3.2):

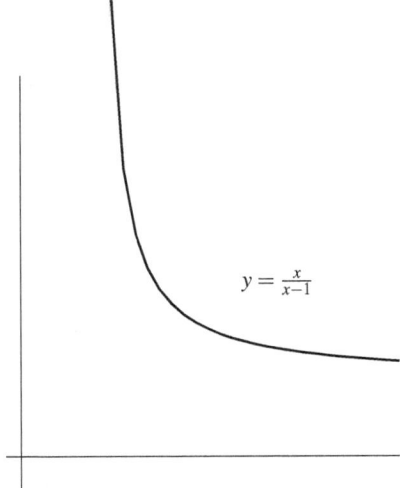

$$y = \tfrac{x}{x-1}$$

Fig. 3.2 Graph of the curve $y = \tfrac{x}{x-1}$.

3.3 Periodic Solutions of Nonautonomous Riccati Difference Equations

Our scheme of this section is to examine the existence, uniqueness, and patterns of periodic cycles of the **Nonautonomous Riccati Difference Equations**:

$$x_{n+1} = \frac{a_n}{x_n} \ , \quad n = 0, 1, \dots , \tag{3.3}$$

and

$$x_{n+1} = \frac{a_n x_n}{x_n - 1} \ , \quad n = 0, 1, \dots , \tag{3.4}$$

where $\{a_n\}_{n=0}^{\infty}$ is a period-k sequence, $(k \geq 2)$. First we will start with Equation (3.3) and assume that $x_0 \neq 0$. It is our goal to demonstrate varieties of combinations of patterns of periodic cycles depending on the period of the period-k sequence $\{a_n\}_{n=0}^{\infty}$ and the relationship of the terms of the period-k sequence. We will show that sometimes the periodic cycles are unique and sometimes every solution is periodic. Now we will exhibit several examples of this phenomenon that we did not encounter in the autonomous case of the previous section.

Example 3. Suppose that $\{a_n\}_{n=0}^{\infty}$ is a period-2 sequence. Show that:

$$x_{n+1} = \frac{a_n}{x_n} \ , \quad n = 0, 1, \dots ,$$

has no period-2 solutions and explain why.

Solution: First we assume that $x_2 = x_0$ and by iteration we get:

$$x_0 \ ,$$

$$x_1 \ = \ \frac{a_0}{x_0} \ ,$$

$$x_2 \ = \ \frac{a_1}{x_1} \ = \ \frac{a_1}{\left[\frac{a_0}{x_0}\right]} \ = \ \frac{a_1 x_0}{a_0} \ = \ x_0 \ .$$

Now we see that $x_2 = x_0$ if and only if $a_0 = a_1$. This is a contradiction as we assumed that $\{a_n\}_{n=0}^{\infty}$ is a period-2 sequence where $a_0 \neq a_1$.

Example 4. Suppose that $\{a_n\}_{n=0}^{\infty}$ is a period-3 sequence. Determine the existence, uniqueness, and pattern of period-3 solutions of:

$$x_{n+1} = \frac{a_n}{x_n} \ , \quad n = 0, 1, \dots .$$

Solution: Assume that $x_3 = x_0$ and we acquire:

$$x_0 \ ,$$

$$x_1 \ = \ \frac{a_0}{x_0} \ ,$$

$$x_2 \ = \ \frac{a_1}{x_1} \ = \ \frac{a_1}{\left[\frac{a_0}{x_0}\right]} \ = \ \frac{a_1 x_0}{a_0} \ ,$$

$$x_3 \ = \ \frac{a_2}{x_2} \ = \ \frac{a_2}{\left[\frac{a_1 x_0}{a_0}\right]} \ = \ \frac{a_2 a_0}{a_1 x_0} \ = \ x_0 \ .$$

Now either $x_0 = \sqrt{\frac{a_2 a_0}{a_1}}$ or $x_0 = -\sqrt{\frac{a_2 a_0}{a_1}}$. Now we will verify the existence of a unique positive period-3 cycle:

$$x_0 = \sqrt{\frac{a_2 a_0}{a_1}}$$

$$x_1 \ = \ \frac{a_0}{x_0} \ = \ \frac{a_0}{\sqrt{\frac{a_2 a_0}{a_1}}} \ = \ \sqrt{\frac{a_0 a_1}{a_2}}$$

$$x_2 \ = \ \frac{a_1}{x_1} \ = \ \frac{a_1}{\sqrt{\frac{a_0 a_1}{a_2}}} \ = \ \sqrt{\frac{a_1 a_2}{a_0}}$$

$$x_3 \ = \ \frac{a_2}{x_2} \ = \ \frac{a_2}{\sqrt{\frac{a_1 a_2}{a_0}}} \ = \ \sqrt{\frac{a_2 a_0}{a_1}} \ = \ x_0 \ .$$

So we see a unique positive period-3 cycle with the following pattern:

$$\sqrt{\frac{a_2 a_0}{a_1}}, \quad \sqrt{\frac{a_0 a_1}{a_2}}, \quad \sqrt{\frac{a_1 a_2}{a_0}}, \quad \ldots \ .$$

Similarly it follows that there is a unique negative period-3 cycle with the following pattern:

$$-\sqrt{\frac{a_2 a_0}{a_1}}, \quad -\sqrt{\frac{a_0 a_1}{a_2}}, \quad -\sqrt{\frac{a_1 a_2}{a_0}}, \quad \ldots \ .$$

Example 5. Suppose that $\{a_n\}_{n=0}^{\infty}$ is a period-4 sequence. Determine the existence, uniqueness, and pattern of period-4 solutions of:

$$x_{n+1} = \frac{a_n}{x_n}, \quad n = 0, 1, \ldots \ .$$

Solution: Similar to the previous examples, set $x_4 = x_0$ and we obtain:

$$x_0 \ ,$$
$$x_1 = \frac{a_0}{x_0},$$
$$x_2 = \frac{a_1}{x_1} = \frac{a_1}{\left[\frac{a_0}{x_0}\right]} = \frac{a_1 x_0}{a_0},$$
$$x_3 = \frac{a_2}{x_2} = \frac{a_2}{\left[\frac{a_1 x_0}{a_0}\right]} = \frac{a_2 a_0}{a_1 x_0},$$
$$x_4 = \frac{a_3}{x_3} = \frac{a_3}{\left[\frac{a_2 a_0}{a_1 x_0}\right]} = \frac{a_3 a_1 x_0}{a_2 a_0} = x_0 \ .$$

We see that $x_4 = x_0$ if and only if $a_3 a_1 = a_2 a_0$. Therefore, every solution is periodic with the following period-4 pattern:

$$x_0 \ , \quad \frac{a_0}{x_0}, \quad \frac{a_1 x_0}{a_0}, \quad \frac{a_2 a_0}{a_1 x_0}, \quad \ldots \ .$$

On one hand, in Example 4, Equation (3.3) has unique periodic cycles when $\{a_n\}_{n=0}^{\infty}$ is an odd ordered period. On the other hand, in Example 5 every solution of Equation (3.3) is periodic when $\{a_n\}_{n=0}^{\infty}$ is an even ordered period. The following two theorems generalize the results.

Theorem 1. *Suppose that $\{a_n\}_{n=0}^{\infty}$ is periodic sequences with period-$(2k+1)$, $(k \in \mathbb{N})$. Then:*

$$x_{n+1} = \frac{a_n}{x_n}, \quad n = 0, 1, \ldots \ ,$$

has a unique period-$(2k+1)$ cycle where:

$$x_0 = \sqrt{\frac{\prod_{i=1}^{k+1} a_{2i-2}}{\prod_{i=1}^{k} a_{2i-1}}} \quad \text{or} \quad x_0 = -\sqrt{\frac{\prod_{i=1}^{k+1} a_{2i-2}}{\prod_{i=1}^{k} a_{2i-1}}} .$$

Proof: *Similar to Example (4), by iteration we get:*

$$x_0 ,$$

$$x_1 = \frac{a_0}{x_0} ,$$

$$x_2 = \frac{a_1}{x_1} = \frac{a_1 x_0}{a_0} ,$$

$$x_3 = \frac{a_2}{x_2} = \frac{a_0 a_2}{a_1 x_0} ,$$

$$x_4 = \frac{a_3}{x_3} = \frac{a_1 a_3 x_0}{a_0 a_2} ,$$

$$x_5 = \frac{a_4}{x_4} = \frac{a_0 a_2 a_4}{a_1 a_3 x_0} ,$$

$$x_6 = \frac{a_5}{x_5} = \frac{a_1 a_3 a_5 x_0}{a_0 a_2 a_4} ,$$

$$x_7 = \frac{a_6}{x_6} = \frac{a_0 a_2 a_4 a_6}{a_1 a_3 a_5 x_0} ,$$

$$\vdots$$

It follows by induction that for all $n \geq 1$:

$$x_{2n+1} = \frac{\prod_{i=1}^{k+1} a_{2i-2}}{\left[\prod_{i=1}^{k} a_{2i-1}\right] x_0} .$$

Thus the result follows.

Theorem 2. *Suppose that $\{a_n\}_{n=0}^{\infty}$ is periodic sequences with period-2k, $(k \in \mathbb{N})$. Then:*

$$x_{n+1} = \frac{a_n}{x_n} , \quad n = 0, 1, \dots ,$$

is periodic with period-2k if and only if:

$$\prod_{i=1}^{k} a_{2i-2} = \prod_{i=1}^{k} a_{2i-1} .$$

The proof of Theorem 2 will be left as an exercise at the end of the chapter. Now we proceed with deciphering the existence and patterns of periodic cycles of Equation (3.4), where $\{a_n\}_{n=0}^{\infty}$ is a period-k sequence, $(k \geq 2)$ and $x_0 \neq 1$. In the next two examples we will encounter unique periodic cycles and determine their patterns when $\{a_n\}_{n=0}^{\infty}$ is an even ordered periodic sequence. The case when $\{a_n\}_{n=0}^{\infty}$ is an odd ordered periodic sequence will be left as a conjecture and exercises at the end of the chapter.

Example 6. Suppose that $\{a_n\}_{n=0}^{\infty}$ is a period-2 sequence. Show that:

$$x_{n+1} = \frac{a_n x_n}{x_n - 1} \quad , \quad n = 0, 1, \dots ,$$

has a unique period-2 cycle.

Solution: Suppose that $x_2 = x_0$ and we get:

$$x_0$$

$$x_1 = \frac{a_0 x_0}{x_0 - 1}$$

$$x_2 = \frac{a_1 [x_1]}{[x_1] - 1} = \frac{a_1 \left[\frac{a_0 x_0}{x_0 - 1} \right]}{\left[\frac{a_0 x_0}{x_0 - 1} \right] - 1} = \frac{a_0 a_1 x_0}{a_0 x_0 - x_0 + 1} = x_0 .$$

Solving for x_0 we get $x_0 = \frac{a_0 a_1 - 1}{a_0 - 1}$, where $a_0 a_1 \neq 1$ and $a_0 \neq 1$. Now by computation and induction we will verify that we have a unique period-2 cycle:

$$x_0 = \frac{a_0 a_1 - 1}{a_0 - 1}$$

$$x_1 = \frac{a_0 [x_0]}{[x_0] - 1} = \frac{a_0 \left[\frac{a_0 a_1 - 1}{a_0 - 1} \right]}{\left[\frac{a_0 a_1 - 1}{a_0 - 1} \right] - 1} = \frac{a_0 a_1 - 1}{a_1 - 1}$$

$$x_2 = \frac{a_1 [x_1]}{[x_1] - 1} = \frac{a_0 \left[\frac{a_0 a_1 - 1}{a_1 - 1} \right]}{\left[\frac{a_0 a_1 - 1}{a_1 - 1} \right] - 1} = \frac{a_0 a_1 - 1}{a_0 - 1} = x_0 .$$

Notice that $a_1 \neq 1$ and we obtain the following unique period-2 pattern:

$$\frac{a_0 a_1 - 1}{a_0 - 1} , \frac{a_0 a_1 - 1}{a_1 - 1} , \frac{a_0 a_1 - 1}{a_0 - 1} , \frac{a_0 a_1 - 1}{a_1 - 1} , \dots .$$

First of all, the pattern of the numerator does not change from term to term; however, the pattern of the denominator changes from term to term. In future examples we will see an opposite phenomenon. Second, compared to Example 2, Equation (3.4) has a unique period-2 cycle.

Example 7. Suppose that $\{a_n\}_{n=0}^{\infty}$ is a period-4 sequence. Show that:

$$x_{n+1} = \frac{a_n x_n}{x_n - 1} \quad , \quad n = 0, 1, \ldots ,$$

has a unique period-4 solution.

Solution: Set $x_4 = x_0$ and we get:

$$x_0$$

$$x_1 = \frac{a_0 x_0}{x_0 - 1}$$

$$x_2 = \frac{a_1 [x_1]}{[x_1] - 1} = \frac{a_1 \left[\frac{a_0 x_0}{x_0 - 1}\right]}{\left[\frac{a_0 x_0}{x_0 - 1}\right] - 1} = \frac{a_0 a_1 x_0}{a_0 x_0 - x_0 + 1}$$

$$x_3 = \frac{a_2 [x_2]}{[x_2] - 1} = \frac{a_2 \left[\frac{a_0 a_1 x_0}{a_0 x_0 - x_0 + 1}\right]}{\left[\frac{a_0 a_1 x_0}{a_0 x_0 - x_0 + 1}\right] - 1} = \frac{a_0 a_1 a_2 x_0}{a_0 a_1 x_0 - a_0 x_0 + x_0 - 1}$$

$$x_4 = \frac{a_3 [x_3]}{[x_3] - 1} = \frac{a_3 \left[\frac{a_0 a_1 a_2 x_0}{a_0 a_1 x_0 - a_0 x_0 + x_0 - 1}\right]}{\left[\frac{a_0 a_1 a_2 x_0}{a_0 a_1 x_0 - a_0 x_0 + x_0 - 1}\right] - 1}$$

$$= \frac{a_0 a_1 a_2 a_3 x_0}{a_0 a_1 a_2 x_0 - a_0 a_1 x_0 + a_0 x_0 - x_0 + 1} .$$

Solving for x_0 gives us

$$x_0 = \frac{a_0 a_1 a_2 a_3 - 1}{a_0 a_1 a_2 - a_0 a_1 + a_0 - 1}$$

where $a_0 a_1 a_2 a_3 \neq 1$ and $a_0 a_1 a_2 - a_0 a_1 + a_0 - 1 \neq 0$ with the following unique period-4 pattern:

$$x_0 = \frac{a_0 a_1 a_2 a_3 - 1}{a_0 a_1 a_2 - a_0 a_1 + a_0 - 1}$$

$$x_1 = \frac{a_0 a_1 a_2 a_3 - 1}{a_1 a_2 a_3 - a_1 a_2 + a_1 - 1}$$

$$x_2 = \frac{a_0 a_1 a_2 a_3 - 1}{a_2 a_3 a_0 - a_2 a_3 + a_2 - 1}$$

$$x_3 = \frac{a_0 a_1 a_2 a_3 - 1}{a_3 a_0 a_1 - a_3 a_0 + a_3 - 1}$$

Similar to the previous example, the pattern of indices in the numerator does not change from neighbor to neighbor; however, the pattern of indices in the denominator does change.

From Example 6 and Example 7, the following theorem describes the result when $\{a_n\}_{n=0}^{\infty}$ is an even ordered periodic sequence.

Theorem 3. *Suppose that $\{a_n\}_{n=0}^{\infty}$ is periodic sequences with period-2k, ($k \in \mathbb{N}$). Then:*

$$x_{n+1} = \frac{a_n x_n}{x_n - 1} \quad , \quad n = 0, 1, \dots ,$$

has a unique period-2k cycle where:

$$x_0 = \frac{\prod_{i=1}^{2k} a_{i-1} - 1}{\prod_{i=1}^{2k-1} a_{i-1} - \prod_{i=1}^{2k-2} a_{i-1} + \dots + a_0 - 1} ,$$

and

$$\prod_{i=1}^{2k} a_{i-1} \neq 1 \quad \text{and} \quad \prod_{i=1}^{2k-1} a_{i-1} - \prod_{i=1}^{2k-2} a_{i-1} + \dots + a_0 - 1 \neq 0 .$$

The proof of Theorem 3 and the case when $\{a_n\}_{n=0}^{\infty}$ is an odd ordered periodic sequence will be left as an exercise at the end of the chapter.

3.4 The Logistic Difference Equation

Our next plan is to analyze the periodic character of the **Logistic Δ.E.** in the form:

$$x_{n+1} = r x_n (1 - x_n) \quad , \quad n = 0, 1, \dots , \tag{3.5}$$

where $r \in (0, 4]$ and $x_0 \in [0, 1)$. Equation (3.5) has many applications in population dynamics, cardiology, and other biological and medical sciences [5, 6]. By setting:

$$\bar{x} = r\bar{x}(1 - \bar{x}) ,$$

we obtain two equilibrium solutions:

$$\bar{x}_1 = 0 \quad \text{and} \quad \bar{x}_2 = \frac{r - 1}{r} \quad \text{(when } r > 1\text{)} .$$

Now we will examine the existence and patterns of periodic cycles. In fact, when $r \in (3, 4]$, then Equation (3.5) has periodic cycles of any period-p ($p \geq 2$). We will introduce periodic traits of Equation (3.5) with some graphical examples (Figures 3.3 and 3.4).

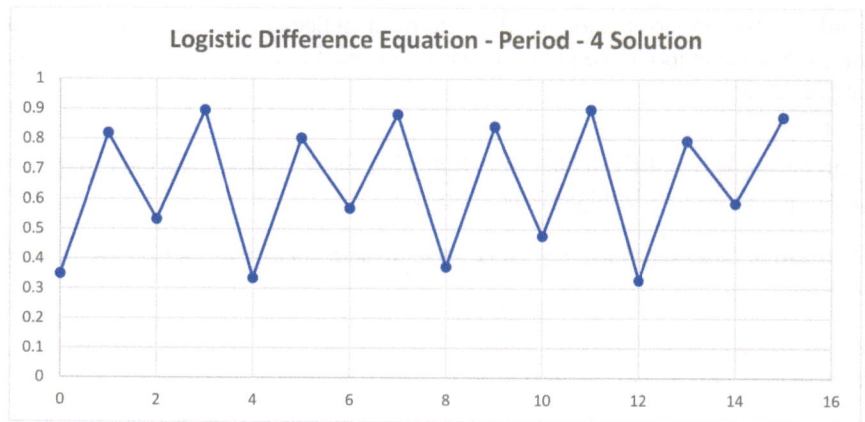

Fig. 3.3 Period-4 cycle with $r = 3.6$ and $x_0 = 0.35$.

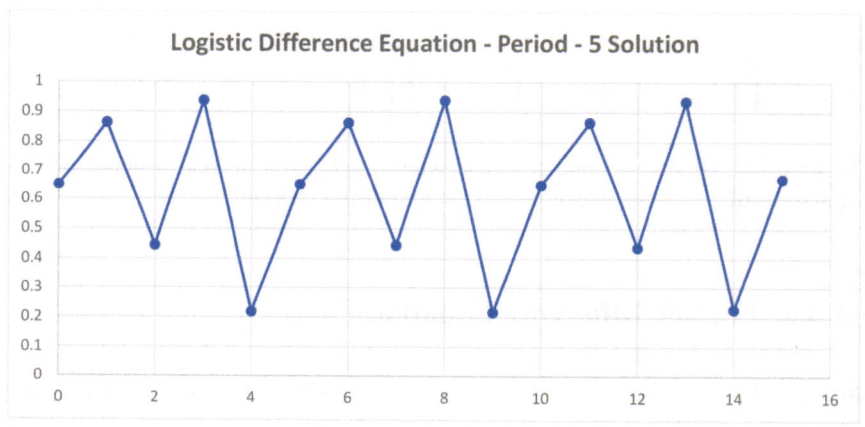

Fig. 3.4 Period-5 cycle with $r = 3.8$ and $x_0 = 0.65$.

Our next objective is to analytically determine the existence and the patterns of periodic solutions. We will start off with the existence and patterns of period-2 solutions.

Example 8. Determine the necessary and sufficient conditions for the existence of period-2 cycles:

$$x_{n+1} = rx_n(1 - x_n) \quad, \quad n = 0, 1, \dots.$$

Solution: Suppose that Equation (3.5) has a minimal period-2 solution,

$$\alpha, \beta, \alpha, \beta, \dots,$$

where $\alpha \neq \beta$ and

$$\alpha = r\beta(1 - \beta) \quad \text{and} \quad \beta = r\alpha(1 - \alpha). \tag{3.6}$$

Now consider the following quadratic equation:

$$(x-\alpha)(x-\beta) = x^2 - x(\alpha+\beta) + \alpha\beta = 0. \tag{3.7}$$

It suffices to determine $\alpha+\beta$ and $\alpha\beta$ in terms of r. Via (3.6) we obtain:

$$\alpha+\beta = r\beta(1-\beta) + r\alpha(1-\alpha) = r(\alpha+\beta) - r(\alpha^2+\beta^2), \tag{3.8}$$

and

$$\alpha-\beta = r\beta(1-\beta) - r\alpha(1-\alpha) = r(\beta-\alpha) - r(\beta^2-\alpha^2), \tag{3.9}$$

As we assumed that $\alpha \neq \beta$, then via (3.9) we procure

$$1 = -r + r(\beta+\alpha). \tag{3.10}$$

Therefore, from (3.10) we derive

$$\alpha+\beta = \frac{1+r}{r}. \tag{3.11}$$

Also, via (3.6) we acquire:

$$\alpha\beta = r^2\alpha\beta(1-\beta)(1-\alpha).$$

Hence

$$1 = r^2(1-\beta)(1-\alpha) = r^2[1-(\alpha+\beta)+\alpha\beta]. \tag{3.12}$$

Then via (3.11) and (3.12):

$$\alpha\beta = \frac{1+r}{r^2}. \tag{3.13}$$

Notice substituting (3.11) and (3.13) into (3.7), we assemble the following quadratic equation:

$$x^2 - x(\alpha+\beta) + \alpha\beta = x^2 - x\left(\frac{1+r}{r}\right) + \frac{1+r}{r^2} = 0. \tag{3.14}$$

Provided that $3 < r \leq 4$, the two distinct solutions of Equation (3.14) are

$$x_1 = \frac{\frac{1+r}{r} + \sqrt{\left(\frac{1+r}{r}\right)^2 - 4\left(\frac{1+r}{r^2}\right)}}{2} \quad \text{and} \quad x_2 = \frac{\frac{1+r}{r} - \sqrt{\left(\frac{1+r}{r}\right)^2 - 4\left(\frac{1+r}{r^2}\right)}}{2},$$

which reduce to

$$x_1 = \frac{(1+r) + \sqrt{(r+1)(r-3)}}{2r} \quad \text{and} \quad x_2 = \frac{(1+r) - \sqrt{(r+1)(r-3)}}{2r}.$$

Since we assumed that $\alpha \neq \beta$, then $r \in (3,4]$. Notice that if $r \leq 3$, then we do not have two district real roots. Hence:

$$\alpha = \frac{(1+r)+\sqrt{(r+1)(r-3)}}{2r} \quad \text{and} \quad \beta = \frac{(1+r)-\sqrt{(r+1)(r-3)}}{2r}.$$

The next two examples will assume that $r = 4$ and will portray the existence of a period-3 cycle and a period-2 cycle.

Example 9. Determine the period of Equation (3.5) by solving the Initial Value Problem:

$$\begin{cases} x_{n+1} = 4x_n(1-x_n), & n = 0, 1, \dots, \\ \\ x_0 = \sin^2\left(\frac{\pi}{7}\right). \end{cases}$$

Solution: By using the double angle identity $\sin(2\theta) = 2\sin(\theta)\cos(\theta)$ and by iteration we get:

$$x_0 = \sin^2\left(\frac{\pi}{7}\right),$$

$$x_1 = 4x_0(1-x_0) = 4\sin^2\left(\frac{\pi}{7}\right)\left[1-\sin^2\left(\frac{\pi}{7}\right)\right]$$
$$= 4\sin^2\left(\frac{2\pi}{7}\right)\cos^2\left(\frac{\pi}{7}\right) = \sin^2\left(\frac{2\pi}{7}\right),$$

$$x_2 = 4x_1(1-x_1) = 4\sin^2\left(\frac{2\pi}{7}\right)\left[1-\sin^2\left(\frac{2\pi}{7}\right)\right] = \sin^2\left(\frac{4\pi}{7}\right),$$

$$x_3 = 4x_2(1-x_2) = 4\sin^2\left(\frac{4\pi}{7}\right)\left[1-\sin^2\left(\frac{4\pi}{7}\right)\right] = \sin^2\left(\frac{8\pi}{7}\right)$$

$$= \sin^2\left(\frac{\pi}{7}\right) = x_0.$$

Thus we acquire the following period-3 pattern:

$$\sin^2\left(\frac{\pi}{7}\right), \ \sin^2\left(\frac{2\pi}{7}\right), \ \sin^2\left(\frac{4\pi}{7}\right), \dots.$$

Example 10. Determine the period of Equation (3.5) by solving the Initial Value Problem:

$$\begin{cases} x_{n+1} = 4x_n(1-x_n), & n = 0, 1, \dots, \\ \\ x_0 = \sin^2\left(\frac{\pi}{5}\right). \end{cases}$$

Solution: Applying the double angle identity $\sin(2\theta) = 2\sin(\theta)\cos(\theta)$ gives us:

$$x_0 = sin^2\left(\frac{\pi}{5}\right),$$

$$x_1 = 4x_0(1-x_0) = 4sin^2\left(\frac{\pi}{5}\right)\left[1-sin^2\left(\frac{\pi}{5}\right)\right] = sin^2\left(\frac{2\pi}{5}\right),$$

$$x_2 = 4x_1(1-x_1) = 4sin^2\left(\frac{2\pi}{5}\right)\left[1-sin^2\left(\frac{2\pi}{5}\right)\right] = sin^2\left(\frac{4\pi}{5}\right) = sin^2\left(\frac{\pi}{5}\right) = x_0.$$

Now we see the following period-2 pattern:

$$sin^2\left(\frac{\pi}{5}\right) , \ sin^2\left(\frac{2\pi}{5}\right) , \ sin^2\left(\frac{\pi}{5}\right) , \ sin^2\left(\frac{2\pi}{5}\right) , \ \dots .$$

From Example 9 and Example 10 we can suggest the following **Open Problem**:

Open Problem 4. *Let $k \in \mathbb{N}$. Then the solution to the Initial Value Problem:*

$$\begin{cases} x_{n+1} = 4x_n(1-x_n) , & n = 0,1,\dots , \\ x_0 = sin^2\left(\frac{2\pi}{2k+1}\right) , \end{cases}$$

is periodic with some period-p, $(p \geq 2)$.

The pertinent question to address: what is the period of Equation (3.5) depending on the value of k? We will leave this question in the end of the chapter exercises. We can generalize that Equation (3.5) has periodic solutions with period-p $(p \geq 2)$ when $r \in (3,4]$. Furthermore, Equation (3.5) has eventually periodic solutions as we can see in the graph below (Figure 3.5):

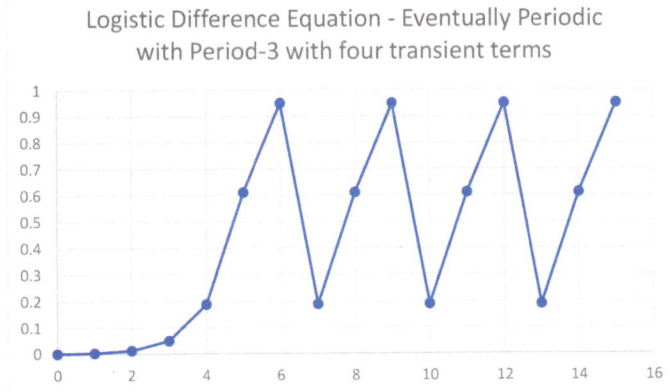

Fig. 3.5 An eventually period-3 cycle with four transient terms with $r = 4$ and $x_0 = sin^2\left(\frac{\pi}{112}\right)$.

More questions on eventually periodic solutions of Equation (3.5) will be addressed at the end of the chapter exercises. In the next section we will examine examples of periodic solutions and eventually periodic solutions of **piece-wise difference equations**.

3.5 The Tent-Map

In this section we will study the existence of periodic solutions and eventually periodic solutions of the **Tent-Map**, which is a specific type of **Piece-Wise Δ.E.** in the form:

$$x_{n+1} = \begin{cases} 2x_n & \text{if } x_n < \frac{1}{2}, \\ \\ 2(1-x_n) & \text{if } x_n \geq \frac{1}{2}. \end{cases} \quad n = 0, 1, \dots, \tag{3.15}$$

where $x_0 \in (0, 1)$. We call these types of difference equations **piece-wise difference equations** as they are composed of two or more pieces defined on a specific interval as piecewise functions are defined on a particular interval. Equation (3.15) has two equilibrium points $\bar{x}_1 = 0$ and $\bar{x}_2 = \frac{2}{3}$. The next three examples will examine the existence, uniqueness, and patterns of the period-2, period-3, and period-4 cycles of Equation (3.15).

Example 11. Determine the period-2 cycle of:

$$x_{n+1} = \begin{cases} 2x_n & \text{if } x_n < \frac{1}{2}, \\ \\ 2(1-x_n) & \text{if } x_n \geq \frac{1}{2}. \end{cases} \quad n = 0, 1, \dots,$$

Solution: Set $x_2 = x_0$ and we acquire:

$$x_0,$$

$$x_1 = 2x_0,$$

$$x_2 = 2(1-x_1) = 2(1-2x_0) = 2 - 4x_0 = x_0.$$

We get $x_0 = \frac{2}{5}$ and the following period-2 pattern (Figure 3.6):

$$\frac{2}{5}, \frac{4}{5}, \frac{2}{5}, \frac{4}{5} \dots$$

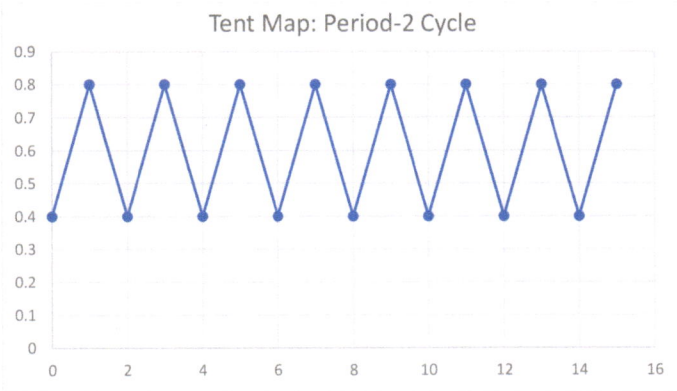

Fig. 3.6 Period-2 cycle with $x_0 = \frac{2}{5}$.

Example 12. Determine a period-3 cycle of:

$$x_{n+1} = \begin{cases} 2x_n & \text{if } x_n < \frac{1}{2}, \\ 2(1-x_n) & \text{if } x_n \geq \frac{1}{2}. \end{cases} \quad n = 0,1,\dots,$$

Solution: Similarly, set $x_3 = x_0$ and we get:

$$x_0,$$

$$x_1 = 2x_0,$$

$$x_2 = 2x_1 = 2[2x_0] = 4x_0,$$

$$x_3 = 2(1-x_2) = 2(1-4x_0) = 2 - 8x_0 = x_0.$$

Notice that $x_0 = \frac{2}{9}$ produces the following period-3 pattern (Figure 3.7):

$$\frac{2}{9}, \frac{4}{9}, \frac{8}{9}, \frac{2}{9}, \frac{4}{9}, \frac{8}{9} \dots$$

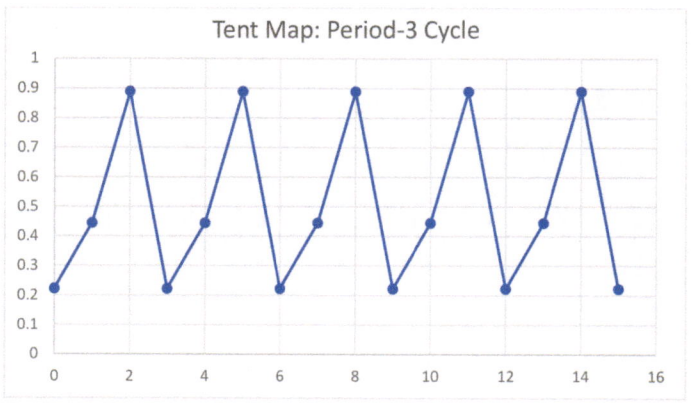

Fig. 3.7 Period-3 cycle with $x_0 = \frac{2}{9}$.

Example 13. Determine a period-4 cycle of:

$$x_{n+1} = \begin{cases} 2x_n & \text{if } x_n < \frac{1}{2}, \\ 2(1-x_n) & \text{if } x_n \geq \frac{1}{2}. \end{cases} \quad n = 0, 1, \dots,$$

Solution: Set $x_4 = x_0$ and we procure:

$$x_0,$$

$$x_1 = 2x_0,$$

$$x_2 = 2x_1 = 2[2x_0] = 4x_0,$$

$$x_3 = 2x_2 = 2[4x_0] = 8x_0,$$

$$x_4 = 2(1-x_3) = 2(1-8x_0) = 2-16x_0 = x_0.$$

Observe that $x_0 = \frac{2}{17}$ and we obtain the following period-4 pattern (Figure 3.8):

$$\frac{2}{17}, \frac{4}{17}, \frac{8}{17}, \frac{16}{17}, \frac{2}{17}, \frac{4}{17}, \frac{8}{17}, \frac{16}{17}, \dots.$$

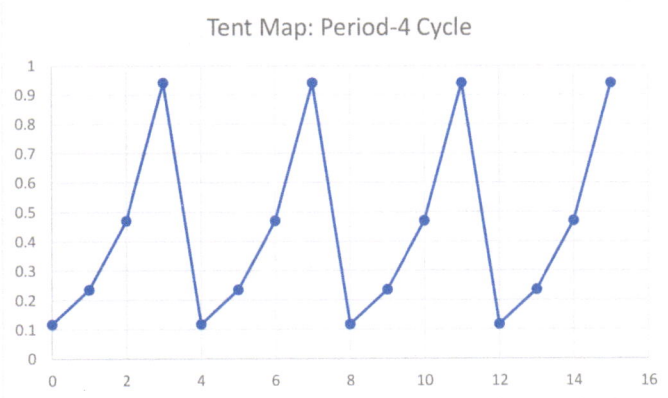

Fig. 3.8 Period-4 cycle with $x_0 = \frac{2}{17}$.

From Examples (11)-(13), we can see that Equation (3.15) has periodic solutions with period-p ($p \geq 2$) and the following theorem describes the result.

Theorem 5. *The* **Tent-Map***:*

$$x_{n+1} = \begin{cases} 2x_n & \text{if } x_n < \frac{1}{2}, \\[2mm] 2(1-x_n) & \text{if } x_n \geq \frac{1}{2}. \end{cases} \quad n = 0, 1, \dots ,$$

has a periodic solution with period-p, ($p \geq 2$), where:

$$x_0 = \frac{2}{2^p + 1} .$$

The proof of Theorem 5 will be left as an exercise at the end of the chapter. From Chapter 1, recall that $\{x_n\}_{n=0}^{\infty}$ is **eventually periodic** with minimal period-p ($p \geq 2$) if there exists $N \in \mathbb{N}$ such that:

$$x_{n+N} = x_{(n+p)+N} \quad \text{for all } n \geq 0 ,$$

In fact, $\{x_n\}_{n=0}^{\infty}$ is eventually periodic with **minimal period-2** if there exists $N \in \mathbb{N}$ such that:

$$x_{n+N} = x_{(n+2)+N} \quad \text{for all } n \geq 0 .$$

The next two examples will analytically and graphically exhibit the existence of N transient terms, ($N \in \mathbb{N}$).

Example 14. Solve the Initial Value Problem:

$$
x_{n+1} = \begin{cases} 2x_n & \text{if } x_n < \frac{1}{2}, \\ 2(1-x_n) & \text{if } x_n \geq \frac{1}{2} \quad\quad n = 0,1,\dots, \\ x_0 = \frac{1}{40}, \end{cases}
$$

and show that the solution is eventually periodic with period-2.

Solution: By iteration we get:

$$x_0 = \frac{1}{40},$$

$$x_1 = 2x_0 = \frac{1}{20},$$

$$x_2 = 2x_1 = \frac{1}{10},$$

$$x_3 = 2x_2 = \frac{1}{5},$$

$$x_4 = 2x_3 = \frac{2}{5},$$

$$x_5 = 2x_4 = \frac{4}{5},$$

$$x_6 = 2(1-x_5) = \frac{2}{5} = x_4.$$

Now notice (Figure 3.9):

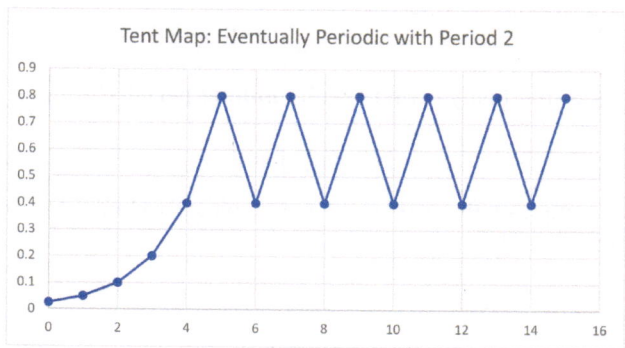

Fig. 3.9 we have four increasing transient terms from $x_0 - x_3$ and $x_6 = x_4$.

Example 15. Solve the Initial Value Problem:

$$x_{n+1} = \begin{cases} 2x_n & \text{if } x_n < \frac{1}{2} , \\ 2(1-x_n) & \text{if } x_n \geq \frac{1}{2} \qquad n = 0, 1, \ldots , \\ x_0 = \frac{1}{288} , \end{cases}$$

and show that the solution is eventually periodic with period-3.

Solution: The graph below (Figure 3.10):

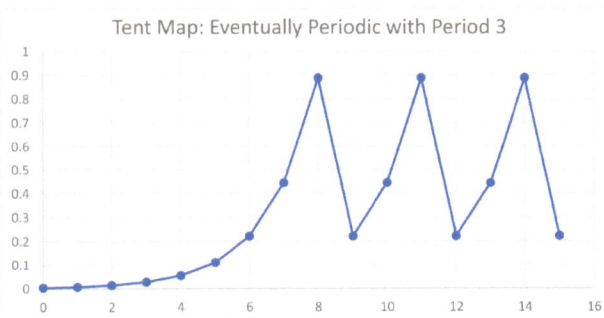

Fig. 3.10 describes six transient terms from $x_0 - x_5$ and $x_9 = x_6$.

Similar to Example 14 and Example 15, we can determine eventually periodic solutions with period-p, $(p \geq 3)$. This will be left as an exercise at the end of the chapter.

3.6 The 3X+1 Conjecture

The **3X+1 Conjecture** also known as the **Collatz Conjecture** was introduced by Lothar Collatz in 1937. Our intent is to explore the periodic solutions and eventually periodic solutions Piece-Wise Δ.E. known as the **3X+1 Conjecture** in the form:

$$x_{n+1} = \begin{cases} \frac{x_n}{2} & \text{if } x_n \text{ is even} , \\ \frac{3x_n+1}{2} & \text{if } x_n \text{ is odd} , \end{cases} \qquad n = 0, 1, \ldots , \qquad (3.16)$$

where $x_0 \in \mathbb{N}$. First of all, $x_0 = 2$ produces the following period-2 pattern:

$$2 , 1 , 2 , 1 , \ldots ,$$

Second, with $x_0 = 1$ we obtain:

$$1, 2, 1, 2, \ldots .$$

Conjecture 1. Let $x_0 \in \mathbb{N}$. Then every solution of:

$$x_{n+1} = \begin{cases} \frac{x_n}{2} & \text{if } x_n \text{ is even}, \\ \\ \frac{3x_n+1}{2} & \text{if } x_n \text{ is odd}, \end{cases} \qquad n = 0, 1, \ldots ,$$

is eventually periodic with the following period-2 cycle:

$$2, 1, 2, 1, \ldots .$$

Now we will show a specific example on how to implement the **3X+1 Conjecture**.

Example 16. Solve the Initial Value Problem:

$$x_{n+1} = \begin{cases} \frac{x_n}{2} & \text{if } x_n \text{ is even}, \\ \\ \frac{3x_n+1}{2} & \text{if } x_n \text{ is odd}, \qquad n = 0, 1, \ldots , \\ \\ x_0 = 17 , \end{cases}$$

and show that the solution is eventually periodic with period-2.

Solution: Notice:

$$x_0 = 17 ,$$

$$x_1 = \frac{3 \cdot 17 + 1}{2} = 26 ,$$

$$x_2 = \frac{26}{2} = 13 ,$$

$$x_3 = \frac{3 \cdot 13 + 1}{2} = 20 ,$$

$$x_4 = \frac{20}{2} = 10 ,$$

$$x_5 = \frac{10}{2} = 5 ,$$

$$x_6 = \frac{3 \cdot 5 + 1}{2} = 8 ,$$

$$x_7 = \frac{8}{2} = 4 \, ,$$

$$x_8 = \frac{4}{2} = 2 \, ,$$

$$x_9 = \frac{2}{2} = 1 \, ,$$

$$x_{10} = \frac{3 \cdot 2 + 1}{2} = 2 \, .$$

Observe that we have eight transient terms from $x_0 - x_7$ and $x_{10} = x_8$:

$$[\, 17 \, , \, 26 \, , \, 13 \, , \, 20 \, , \, 10 \, , \, 5 \, , \, 8 \, , \, 4 \,] \, , \, 2 \, , \, 1 \, , 2 \, , \, 1 \dots \, .$$

The transient terms are emphasized in square brackets.

The **3X+1 Conjecture** claims that every solution of Equation (3.16) is eventually periodic with period-2. However, only special cases have been proved. We will advance our studies with periodic nature of solutions of another Piece-Wise Δ.E. that describes a neural networking model.

3.7 Autonomous Piece-wise Difference Equation as a Neuron Model

In this section, our plan is to study the existence of periodic solutions and eventually periodic solutions of the Autonomous Piece-wise Δ.E. (discrete-time network of a single neuron model) in the form:

$$x_{n+1} = \beta x_n - g(x_n) \, , \quad n = 0, 1, 2, \dots \, , \tag{3.17}$$

where $x_0 \in \Re$, $\beta > 0$ is an internal decay rate, and g is a signal function. Equation (3.17) is examined in several articles [3, 4, 11, 10] and is analyzed as a single neuron model, where a signal function g is the following piece-wise constant with the **McCulloch-Pitts nonlinearity**:

$$g(x) = \begin{cases} 1 & \text{if } x \geq 0 \, , \\ -1 & \text{if } x < 0 \, . \end{cases}$$

Piece-wise difference equations have been used as mathematical models for various applications, including neurons [36, 37, 38, 39, 40]. We will manifest several examples that exhibit periodic cycles with various periods of Equation (3.17).

Example 17. Determine the period-2 cycle of:

$$x_{n+1} = \beta x_n - g(x_n) \, , \quad n = 0, 1, 2, \dots \, ,$$

where $x_0 \in \mathfrak{R}$, $\beta > 0$ and

$$g(x) = \begin{cases} 1 & \text{if } x \geq 0, \\ -1 & \text{if } x < 0. \end{cases}$$

Solution: Assuming that $0 < x_0 < \frac{1}{\beta}$ and setting $x_2 = x_0$ give us:

$$x_0 > 0,$$

$$x_1 = \beta x_0 - 1,$$

$$x_2 = \beta [x_1] + 1 = \beta [\beta x_0 - 1] + 1 = \beta^2 x_0 - \beta + 1 = x_0.$$

Thus $x_0 = \frac{1}{\beta+1}$ and generates the following period-2 pattern:

$$\frac{1}{\beta+1}, \quad -\frac{1}{\beta+1}, \quad \frac{1}{\beta+1}, \quad -\frac{1}{\beta+1}, \quad \dots.$$

Notice that two neighboring terms are of opposite sign. In addition we get (Figure 3.11):

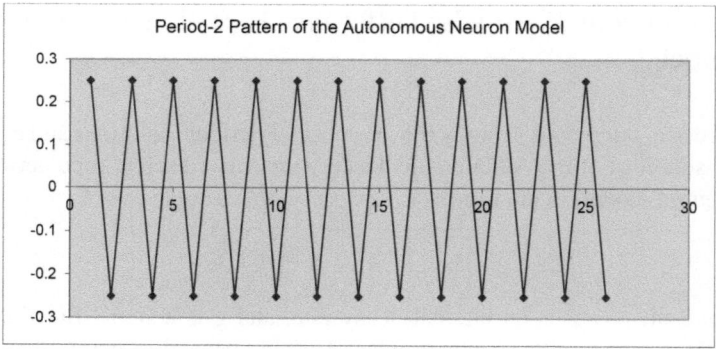

Fig. 3.11 a period-2 cycle with $\beta = 3$ and $x_0 = \frac{1}{4}$. Note that the sum of two neighboring terms is always 0.

Example 18. Determine a period-3 cycle of:

$$x_{n+1} = \beta x_n - g(x_n), \quad n = 0, 1, 2, \dots,$$

where $x_0 \in \mathfrak{R}$, $\beta > 0$, and

$$g(x) = \begin{cases} 1 & \text{if } x \geq 0, \\ -1 & \text{if } x < 0. \end{cases}$$

Solution: Suppose $\frac{1}{\beta} < x_0 < \frac{\beta+1}{\beta^2}$ and $x_3 = x_0$. Then:

$$x_0 > 0 \, ,$$

$$x_1 = \beta x_0 - 1 \, ,$$

$$x_2 = \beta [x_1] - 1 = \beta [\beta x_0 - 1] - 1 = \beta^2 x_0 - \beta - 1 \, ,$$

$$x_3 = \beta [x_2] + 1 = \beta [\beta^2 x_0 - \beta - 1] + 1 = \beta^3 x_0 - \beta^2 - \beta + 1 = x_0 \, .$$

Therefore $x_0 = \frac{\beta^2 + \beta - 1}{\beta^3 - 1}$ presents the following period-3 pattern:

$$\frac{\beta^2 + \beta - 1}{\beta^3 - 1} \, , \ \frac{\beta^2 - \beta + 1}{\beta^3 - 1} \, , \ \frac{-\beta^2 + \beta + 1}{\beta^3 - 1} \, , \ \dots \dots \, ,$$

Notice that the sign of the $\beta^2, \beta, 1$ in the numerator switches from neighbor to neighbor and:

$$\left[\frac{\beta^2 + \beta - 1}{\beta^3 - 1} \right] + \left[\frac{\beta^2 - \beta + 1}{\beta^3 - 1} \right] + \left[\frac{-\beta^2 + \beta + 1}{\beta^3 - 1} \right]$$
$$= \frac{\beta^2 + \beta + 1}{(\beta - 1)(\beta^2 + \beta + 1)} = \frac{1}{\beta - 1} = \bar{x} \, .$$

Therefore period-3 cycles exist provided that $\beta \neq 1$ (this is not the only period-3 cycle). Now observe (Figure 3.12):

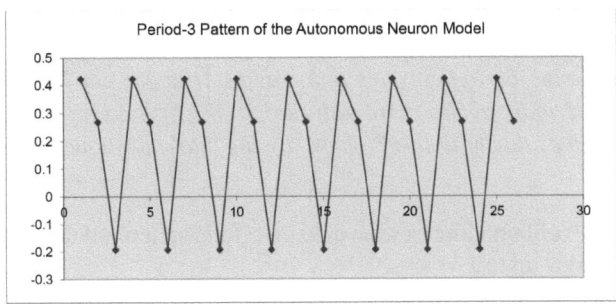

Fig. 3.12 a period-3 cycle where two terms of the cycle are positive and one term of the cycle is negative; $\beta = 3$ and $x_0 = \frac{11}{26}$.

The graph below describes (Figure 3.13):

From Examples 17 and 18 we can determine the patterns of even ordered cycles and odd ordered cycles of Equation (3.17). Furthermore, should all the terms of the even ordered cycles of Equation (3.17) sum up to 0? The following **Open Problems** address these two questions.

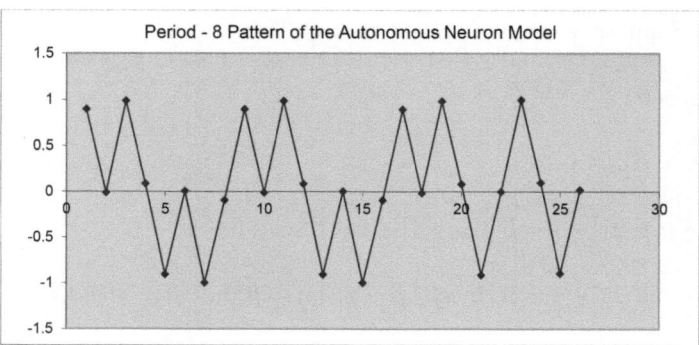

Fig. 3.13 a period-8 cycle with $\beta = 1.1$ and $x_0 = 0.9$.

Open Problem 6. *The $\Delta.E.$:*

$$x_{n+1} = \beta x_n - g(x_n) \ , \quad n = 0, 1, 2, \ldots ,$$

where $x_0 \in \Re$, $\beta \neq 1$, and

$$g(x) = \begin{cases} 1 \ \text{if } x \geq 0 , \\ -1 \ \text{if } x < 0 , \end{cases}$$

has a period-2k solution, ($k \in \mathbb{N}$) and

$$\sum_{i=1}^{2k} x_{i-1} = 0 .$$

Notice that for some periodic cycles this may be true but not for others. First of all, how many periodic cycles exist with period-2k? Second, for which particular periodic cycles does this hold true? These are the vital questions to consider investigating.

The next **Open Problem** addresses the question if the sum of all the terms of the odd ordered cycles adds up to $\bar{x} = \frac{1}{\beta-1}$?

Open Problem 7. *The $\Delta.E.$:*

$$x_{n+1} = \beta x_n - g(x_n) \ , \quad n = 0, 1, 2, \ldots ,$$

where $x_0 \in \Re$, $\beta \neq 1$, and

$$g(x) = \begin{cases} 1 \ \text{if } x \geq 0 , \\ -1 \ \text{if } x < 0 , \end{cases}$$

has a period-$(2k+1)$ solution, $(k \in \mathbb{N})$ and

$$\sum_{i=1}^{2k+1} x_{i-1} = \bar{x} = \frac{1}{\beta-1} \, .$$

Notice that for some periodic cycles this may be true but not for others. How many periodic cycles exist with period-$(2k+1)$? For which particular periodic cycles does this hold true?

Now we will proceed with determining eventually periodic solutions with period-2 of Equation (3.17). The next two examples will show how to procure N transient terms analytically, $(N \in \mathbb{N})$.

Example 19. Determine the period-2 solution with two transient terms of:

$$x_{n+1} = \beta x_n - g(x_n) \, , \quad n = 0, 1, 2, \ldots \, ,$$

where $x_0 \in \mathfrak{R}$, $\beta \neq 1$, and

$$g(x) = \begin{cases} 1 & \text{if } x \geq 0 \, , \\ -1 & \text{if } x < 0 \, . \end{cases}$$

Solution: Set $x_4 = x_2$ and suppose that $\frac{1+\beta+\beta^2}{\beta^3} < x_0 < \frac{1+\beta+\beta^2+\beta^3}{\beta^4}$. Then:

$$x_0 > 0 \, ,$$

$$x_1 = \beta x_0 - 1 \, ,$$

$$x_2 = \beta [x_1] - 1 = \beta [\beta x_0 - 1] - 1 = \beta^2 x_0 - \beta - 1 \, ,$$

$$x_3 = \beta [x_2] - 1 = \beta [\beta^2 x_0 - \beta - 1] - 1 = \beta^3 x_0 - \beta^2 - \beta - 1 \, ,$$

$$x_4 = \beta [x_3] + 1 = \beta [\beta^3 x_0 - \beta^2 - \beta - 1] + 1$$

$$= \beta^4 x_0 - \beta^3 - \beta^2 - \beta + 1 = x_2 \, .$$

By solving for $x_0 = \frac{\beta^2(\beta+1)-2}{\beta^2(\beta^2-1)}$ we obtain:

$$x_0 = \frac{\beta^2(\beta+1)-2}{\beta^2(\beta^2-1)} \, ,$$

$$x_1 = \beta\,[x_0] - 1 = \beta\left[\frac{\beta^2(\beta+1)-2}{\beta^2(\beta^2-1)}\right] - 1 = \frac{\beta+2}{\beta(\beta+1)} \, ,$$

$$x_2 = \beta\,[x_1] - 1 = \beta\left[\frac{\beta+2}{\beta(\beta+1)}\right] - 1 = \frac{1}{\beta+1} \, ,$$

$$x_3 = \beta\,[x_2] - 1 = \beta\left[\frac{1}{\beta+1}\right] - 1 = \frac{-1}{\beta+1} \, ,$$

$$x_4 = \beta\,[x_3] + 1 = \beta\left[\frac{-1}{\beta+1}\right] + 1 = \frac{1}{\beta+1} = x_2 \, ,$$

and generate the following pattern with two transient terms:

$$\left[\frac{\beta^2(\beta+1)-2}{\beta^2(\beta^2-1)} \, , \, \frac{\beta+2}{\beta(\beta+1)}\right] \, , \, \frac{1}{\beta+1} \, , \frac{-1}{\beta+1} \, , \dots \, .$$

Furthermore we see (Figure 3.14):

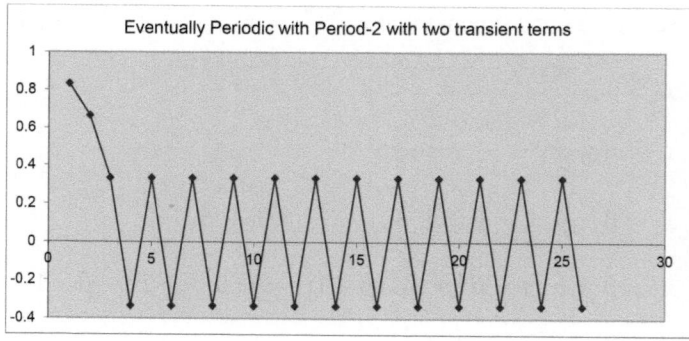

Fig. 3.14 an eventually periodic cycle with period-2 with two decreasing transient terms; $\beta = 2$ and $x_0 = \frac{5}{6}$.

Example 20. Determine the period-2 solution with four transient terms of:

$$x_{n+1} = \beta x_n - g(x_n) \, , \quad n = 0, 1, 2, \dots \, ,$$

where $x_0 \in \mathfrak{R}$, $\beta \neq 1$, and

$$g(x) = \begin{cases} 1 & \text{if } x \geq 0 \, , \\ -1 & \text{if } x < 0 \, . \end{cases}$$

Solution: Set $x_6 = x_4$ and suppose that $\frac{1+\beta+\beta^2+\beta^3+\beta^4}{\beta^5} < x_0 < \frac{1+\beta+\beta^2+\beta^3+\beta^4+\beta^5}{\beta^6}$.
Then:

$$x_0 > 0 \,,$$

$$x_1 = \beta x_0 - 1 \,,$$

$$x_2 = \beta\,[x_1] - 1 = \beta\,[\beta x_0 - 1] - 1 = \beta^2 x_0 - \beta - 1 \,,$$

$$x_3 = \beta\,[x_2] - 1 = \beta\,[\beta^2 x_0 - \beta - 1] - 1 = \beta^3 x_0 - \beta^2 - \beta - 1 \,,$$

$$x_4 = \beta\,[x_3] - 1 = \beta\,[\beta^3 x_0 - \beta^2 - \beta - 1] - 1 = \beta^4 x_0 - \beta^3 - \beta^2 - \beta - 1 \,,$$

$$x_5 = \beta\,[x_4] - 1 = \beta\,[\beta^4 x_0 - \beta^3 - \beta^2 - \beta - 1] - 1$$

$$\quad = \beta^5 x_0 - \beta^4 - \beta^3 - \beta^2 - \beta - 1 \,,$$

$$x_6 = \beta\,[x_5] + 1 = \beta\left[\beta^5 x_0 - \beta^4 - \beta^3 - \beta^2 - \beta - 1\right] + 1$$

$$\quad = \beta^6 x_0 - \beta^5 - \beta^4 - \beta^3 - \beta^2 - \beta + 1 = x_4 \,.$$

Notice $x_0 = \frac{\beta^4(\beta+1)-2}{\beta^4(\beta^2-1)}$ produces the following pattern with four transient terms:

$$\left[\frac{\beta^4(\beta+1)-2}{\beta^4(\beta^2-1)} \,,\; \frac{\beta^3(\beta+1)-2}{\beta^3(\beta^2-1)} \,,\; \frac{\beta^2(\beta+1)-2}{\beta^2(\beta^2-1)} \,,\; \frac{\beta+2}{\beta(\beta+1)}\right] \,,\; \frac{1}{\beta+1} \,,\; \frac{-1}{\beta+1} \,,\; \dots \,.$$

Now we see (Figure 3.15):

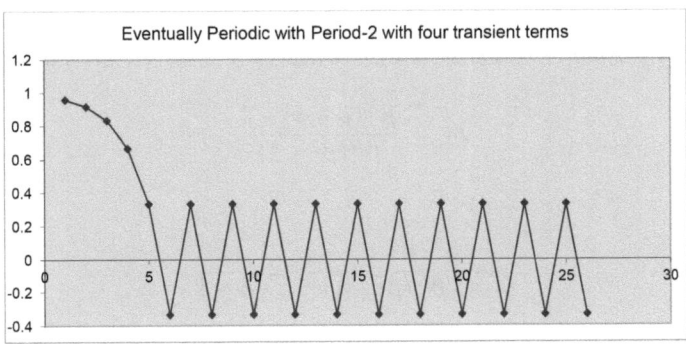

Fig. 3.15 an eventually periodic cycle with period-2 with four decreasing transient terms; $\beta = 2$ and $x_0 = \frac{23}{24}$. This similar phenomenon occurred in Example 19.

The next graph describes (Figure 3.16):

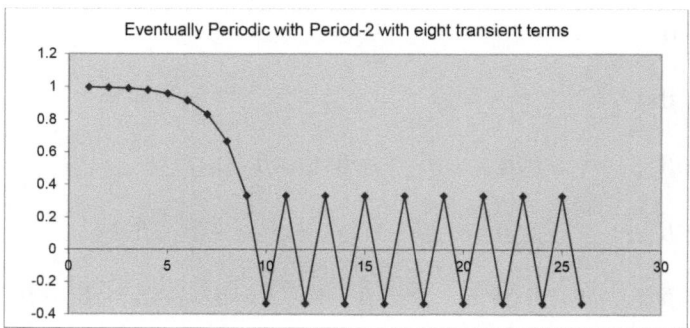

Fig. 3.16 an eventually periodic cycle with period-2 with eight decreasing transient terms; $\beta = 2$ and $x_0 = \frac{383}{384}$.

From Examples 19 and 20 we can inductively determine the general pattern of the N transient terms ($N \in \mathbb{N}$). The following theorem extends the result and will be left as an exercise to prove.

Theorem 8. *The $\Delta.E$.*

$$x_{n+1} = \beta x_n - g(x_n) \ , \ n = 0,1,2, \ldots ,$$

where $x_0 \in \mathfrak{R}$, $\beta \neq 1$, and

$$g(x) = \begin{cases} 1 & \text{if } x \geq 0 , \\ -1 & \text{if } x < 0 , \end{cases}$$

is eventually periodic with period-2 with N-transient terms, ($N \in \mathbb{N}$) when:

$$x_0 = \frac{\beta^N(\beta+1) - 2}{\beta^N(\beta^2 - 1)} .$$

Then

$$\lim_{N \to \infty} \frac{\beta^N(\beta+1) - 2}{\beta^N(\beta^2 - 1)} = \frac{1}{\beta - 1} .$$

Similar to Examples 19 and 20, we can determine the existence of eventually periodic solutions with period-p, ($p \geq 3$). This will be left as an exercise at the end of the chapter.

3.8 Autonomous Piece-wise Difference Equation as a Neuron Model when $\beta = 1$

In this section our intent is to study the patterns of periodic solutions and eventually periodic solutions of the Autonomous Piece-wise Δ.E.:

$$x_{n+1} = x_n - g(x_n) \ , \ \ n = 0, 1, 2, \ldots , \tag{3.18}$$

where $x_0 \in \Re$ and

$$g(x) = \begin{cases} 1 \ \text{if} \ x \geq 0 , \\ -1 \ \text{if} \ x < 0 . \end{cases}$$

Our intent is to show that Equation (3.18) only has period-2 cycles. First of all, we will show that every solution Equation (3.18) is either periodic with period-2 or eventually periodic with period-2. Second, we will not encounter unique period-2 solutions as we did with Equation (3.17).

Example 21. Determine a period-2 solution of:

$$x_{n+1} = x_n - g(x_n) \ , \ \ n = 0, 1, 2, \ldots ,$$

where $x_0 \in \Re$ and

$$g(x) = \begin{cases} 1 \ \text{if} \ x \geq 0 , \\ -1 \ \text{if} \ x < 0 . \end{cases}$$

Solution: Suppose $0 < x_0 < 1$. Then:

$$x_0 > 0 ,$$

$$x_1 = x_0 - 1 < 0 \ \ (\text{as} \ 0 < x_0 < 1) ,$$

$$x_2 = [x_1] + 1 = [x_0 - 1] + 1 = x_0 .$$

If $0 < x_0 < 1$, then every solution of Equation (3.18) is periodic with period-2. In addition, if $x_0 = 0$, then we produce the following period-2 pattern:

$$0 , \ -1 , \ 0 , \ -1 , \ \ldots .$$

Similar pattern can be acquired by letting $x_0 = -1$. Now suppose $x_0 = 1$. Then (Figure 3.17):

The immediate question to ask: is every solution either periodic or eventually periodic? The answer is yes and we will demonstrate several examples.

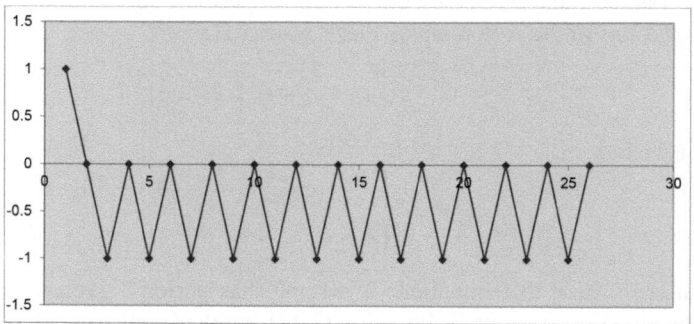

Fig. 3.17 Equation (3.18) is eventually periodic with period-2 with one transient term.

Example 22. Determine the pattern of eventually periodic solution of:

$$x_{n+1} = x_n - g(x_n) \ , \ n = 0, 1, 2, \ \ldots \ ,$$

with $x_0 = 8$ and

$$g(x) = \begin{cases} 1 \ \text{if} \ x \geq 0 \,, \\ -1 \ \text{if} \ x < 0 \,. \end{cases}$$

Solution: The graph below describes (Figure 3.18):

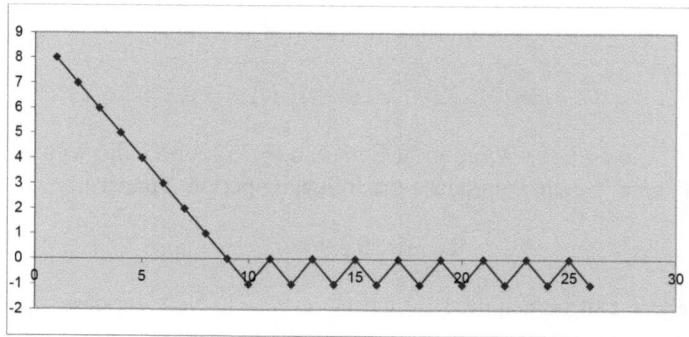

Fig. 3.18 eight positive decreasing transient terms. Furthermore, the transient terms are decreasing linearly in increments of one.

Example 23. Determine the pattern of eventually periodic solution of:

$$x_{n+1} = x_n - g(x_n) \ , \quad n = 0, 1, 2, \ldots \ ,$$

with $x_0 = -7$ and

$$g(x) = \begin{cases} 1 & \text{if } x \geq 0, \\ -1 & \text{if } x < 0. \end{cases}$$

Solution: By evaluating the graph below (Figure 3.19):

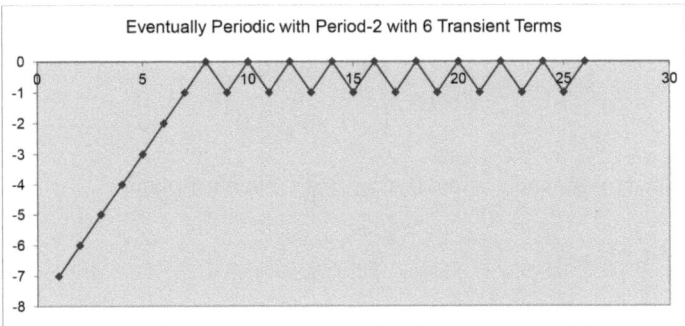

Fig. 3.19 Similar to the previous example, we see exactly six negative transient terms that are strictly increasing. Furthermore, the transient terms are increasing linearly in increments of one.

From Examples (21)–(23), it follows that if x_0 is a positive integer $N \in \mathbb{N}$, then we will have N transient terms that are all decreasing in increments of 1. Similarly, if we let $x_0 = N$, where N is a negative integer, then we will have N negative increasing transient terms increasing in increments of 1. The transient terms of Equation (3.18) either increase or decrease linearly compared to the pattern of transient terms in the previous section when $\beta \neq 1$.

3.9 Nonautonomous Piece-wise Difference Equation as a Neuron Model

In this section we will study the periodic solutions and eventually periodic solutions of the Nonautonomous Piece-wise Δ.E.:

$$x_{n+1} = \beta_n x_n - g(x_n) \ , \quad n = 0, 1, 2, \ldots \ , \tag{3.19}$$

where $x_0 \in \mathfrak{R}$, $\{\beta_n\}_{n=0}^{\infty}$ is a period-2 sequence, and

$$g(x) = \begin{cases} 1 & \text{if } x \geq 0 , \\ -1 & \text{if } x < 0 . \end{cases}$$

We will show that Equation (3.19) only has even ordered period cycles with period-2k, ($k \in \mathbb{N}$). The next two examples will determine the existence of period-2 cycles and period-4 cycles of Equation (3.19).

Example 24. Determine the period-2 cycle of:

$$x_{n+1} = \beta_n x_n - g(x_n) , \quad n = 0, 1, 2, \ldots ,$$

where $x_0 \in \mathbb{R}$, $\{\beta_n\}_{n=0}^{\infty}$ is a period-2 sequence, and

$$g(x) = \begin{cases} 1 & \text{if } x \geq 0 , \\ -1 & \text{if } x < 0 . \end{cases}$$

Solution: Set $x_2 = x_0$ and assume $0 < x_0 < \frac{1}{\beta_0}$. Then we obtain:

$$x_0 > 0 ,$$

$$x_1 = \beta_0 x_0 - 1 ,$$

$$x_2 = \beta_1 [x_1] + 1 = \beta_1 [\beta_0 x_0 - 1] + 1 = \beta_0 \beta_1 x_0 - \beta_1 + 1 = x_0 .$$

We see that $x_0 = \frac{\beta_1 - 1}{\beta_0 \beta_1 - 1}$ and we acquire the following period-2 pattern:

$$x_0 = \frac{\beta_1 - 1}{\beta_0 \beta_1 - 1} ,$$

$$x_1 = \beta_0 [x_0] - 1 = \beta_0 \left[\frac{\beta_1 - 1}{\beta_0 \beta_1 - 1} \right] - 1 = \frac{1 - \beta_0}{\beta_0 \beta_1 - 1} ,$$

$$x_2 = \beta_1 [x_1] + 1 = \beta_1 \left[\frac{1 - \beta_0}{\beta_0 \beta_1 - 1} \right] + 1 = \frac{\beta_1 - 1}{\beta_0 \beta_1 - 1} = x_0 ,$$

First of all, notice that $\beta_0 \beta_1 \neq 1$. Second, note that the two neighboring terms of the period-2 cycle have the same denominator and the indices in the numerator and the signs change from term to term. The following graph describes (Figure 3.20):

Now observe that two neighboring terms are of opposite sign as we saw in the autonomous case in Example 17. However, the sum of the neighboring terms does not add up to 0:

$$\left[\frac{\beta_1 - 1}{\beta_0 \beta_1 - 1} \right] + \left[\frac{1 - \beta_0}{\beta_0 \beta_1 - 1} \right] = \frac{\beta_1 - \beta_0}{\beta_0 \beta_1 - 1} ,$$

as we assumed that $\beta_0 \neq \beta_1$.

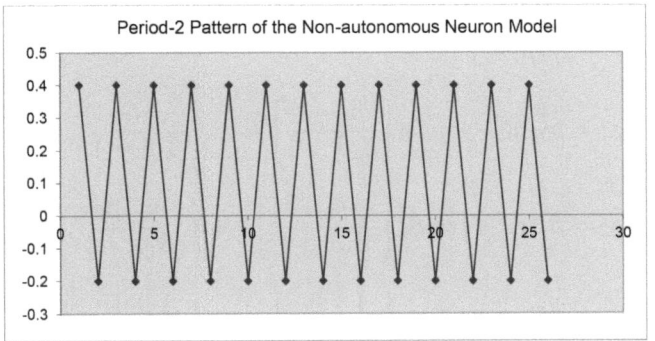

Fig. 3.20 an alternating period-2 pattern with $\beta_0 = 2$, $\beta_1 = 3$, and $x_0 = \frac{2}{5}$.

Example 25. Determine the period-4 solution of:

$$x_{n+1} = \beta_n x_n - g(x_n) \ , \quad n = 0, 1, 2, \ldots ,$$

where $x_0 \in \Re$, $\{\beta_n\}_{n=0}^{\infty}$ is a period-2 sequence, and

$$g(x) = \begin{cases} 1 & \text{if } x \geq 0 , \\ -1 & \text{if } x < 0 . \end{cases}$$

Solution: Set $x_4 = x_0$ and assume $0 < x_0 < \frac{1}{\beta_0}$. By iteration we get:

$$x_0 > 0 ,$$

$$x_1 = \beta_0 x_0 - 1 ,$$

$$x_2 = \beta_1 [x_1] + 1 = \beta_1 [\beta_0 x_0 - 1] + 1 = \beta_0 \beta_1 x_0 - \beta_1 + 1 ,$$

$$x_3 = \beta_0 [x_2] + 1 = \beta_0 [\beta_0 \beta_1 x_0 - \beta_1 + 1] + 1$$

$$= \beta_0^2 \beta_1 x_0 - \beta_0 \beta_1 + \beta_0 + 1 ,$$

$$x_4 = \beta_1 [x_3] - 1 = \beta_1 \left[\beta_0^2 \beta_1 x_0 - \beta_0 \beta_1 + \beta_0 + 1 \right] - 1$$

$$= \beta_0^2 \beta_1^2 x_0 - \beta_0 \beta_1^2 + \beta_0 \beta_1 - 1 = x_0 .$$

Notice $x_0 = \frac{\beta_1 - 1}{\beta_0 \beta_1 + 1}$, and we acquire the following alternating period-4 pattern:

$$x_0 = \frac{\beta_1 - 1}{\beta_0 \beta_1 + 1} \ ,$$

$$x_1 = \beta_0 [x_0] - 1 = \beta_0 \left[\frac{\beta_1 - 1}{\beta_0 \beta_1 + 1} \right] - 1 = \frac{-(\beta_0 + 1)}{\beta_0 \beta_1 + 1} \ ,$$

$$x_2 = \beta_1 [x_1] + 1 = \beta_1 \left[\frac{-(\beta_0 + 1)}{\beta_0 \beta_1 + 1} \right] + 1 = \frac{1 - \beta_1}{\beta_0 \beta_1 + 1} \ ,$$

$$x_3 = \beta_0 [x_2] + 1 = \beta_0 \left[\frac{1 - \beta_1}{\beta_0 \beta_1 + 1} \right] + 1 = \frac{\beta_0 + 1}{\beta_0 \beta_1 + 1} \ ,$$

$$x_4 = \beta_1 [x_3] - 1 = \beta_1 \left[\frac{\beta_0 + 1}{\beta_0 \beta_1 + 1} \right] - 1 = \frac{\beta_1 - 1}{\beta_0 \beta_1 + 1} = x_0 \ ,$$

Notice that all the terms of the period-4 cycle have the same denominator and $x_0 = -x_2$ and $x_1 = -x_3$. This is not the only period-4 cycle. The following graph depicts (Figure 3.21):

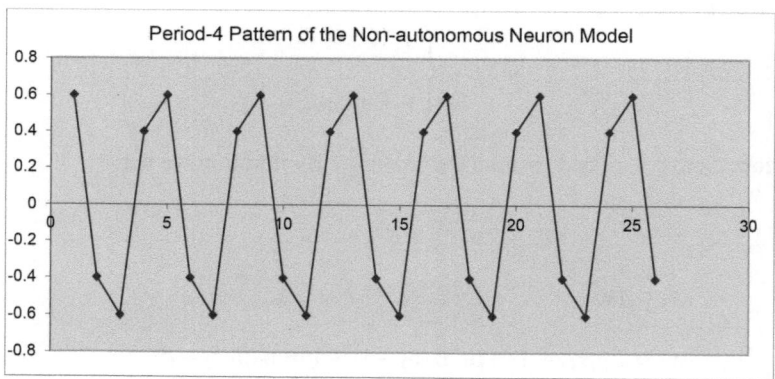

Fig. 3.21 an alternating period-4 pattern with $\beta_0 = 1$, $\beta_1 = 4$, and $x_0 = \frac{3}{5}$.

Furthermore, the sum of all the terms add up to 0 as we can see below:

$$\left[\frac{\beta_1 - 1}{\beta_0 \beta_1 + 1} \right] + \left[\frac{-(\beta_0 + 1)}{\beta_0 \beta_1 + 1} \right] + \left[\frac{1 - \beta_1}{\beta_0 \beta_1 + 1} \right] + \left[\frac{\beta_0 + 1}{\beta_0 \beta_1 + 1} \right] = 0 \ .$$

From Examples 24 and 25 we can inductively determine the pattern of even ordered periods and we can prove that there are no odd ordered periods. This will be left as an exercise. The following graph is an example of (Figure 3.22):

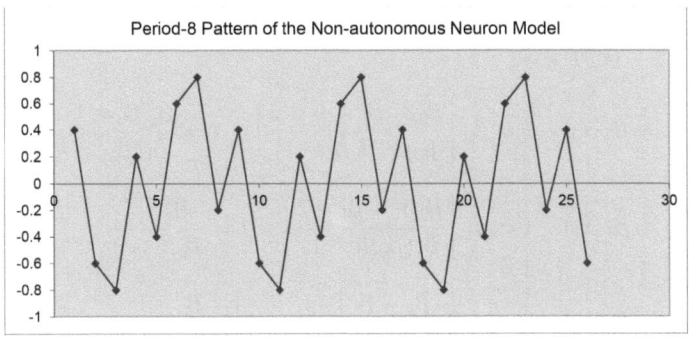

Fig. 3.22 an alternating period-8 pattern with $\beta_0 = 2$, $\beta_1 = 3$, and $x_0 = 0.4$.

Now we will shift our focus on eventually periodic solutions with period-2 of Equation (3.19) with 2N transient terms, ($N \in \mathbb{N}$). Equation (3.19) have even number of transient terms.

Example 26. Determine the period-2 solution with two transient terms of:

$$x_{n+1} = \beta_n x_n - g(x_n) \ , \ \ n = 0, 1, 2, \ldots ,$$

where $x_0 \in \mathfrak{R}$, $\{\beta_n\}_{n=0}^{\infty}$ is a period-2 sequence, and

$$g(x) = \begin{cases} 1 & \text{if } x \geq 0 , \\ -1 & \text{if } x < 0 . \end{cases}$$

Solution: Assume that $x_4 = x_2$, $\beta_0 \beta_1 > 1$, and $\frac{\beta_1 + 1}{\beta_0 \beta_1} < x_0 < \frac{\beta_0[\beta_1 + 1] + 1}{\beta_0[\beta_0 \beta_1]}$. Then:

$$x_0 > 0 ,$$

$$x_1 = \beta_0 x_0 - 1 ,$$

$$x_2 = \beta_1 [x_1] - 1 = \beta_1 [\beta_0 x_0 - 1] - 1 = \beta_0 \beta_1 x_0 - \beta_1 - 1 ,$$

$$x_3 = \beta_0 [x_2] - 1 = \beta_0 [\beta_0 \beta_1 x_0 - \beta_1 - 1] - 1$$

$$= \beta_0^2 \beta_1 x_0 - \beta_0 \beta_1 - \beta_0 - 1 ,$$

$$x_4 = \beta_1 [x_3] + 1 = \beta_1 [\beta_0^2 \beta_1 x_0 - \beta_0 \beta_1 - \beta_0 - 1] + 1$$

$$= \beta_0^2 \beta_1^2 x_0 - \beta_0 \beta_1^2 - \beta_0 \beta_1 - \beta_1 + 1 = x_2 .$$

Hence $x_0 = \frac{\beta_0 \beta_1 [\beta_1 + 1] - 2}{\beta_0 \beta_1 (\beta_0 \beta_1 - 1)}$ and $\beta_0 \beta_1 > 1$. Notice that:

$$x_0 = \frac{\beta_0\beta_1[\beta_1+1]-2}{\beta_0\beta_1(\beta_0\beta_1-1)} \, ,$$

$$x_1 = \beta_0[x_0]-1 = \beta_0\left[\frac{\beta_0\beta_1[\beta_1+1]-2}{\beta_0\beta_1(\beta_0\beta_1-1)}\right]-1 = \frac{\beta_1[\beta_0+1]-2}{\beta_1(\beta_0\beta_1-1)} \, ,$$

$$x_2 = \beta_1[x_1]-1 = \beta_1\left[\frac{\beta_0\beta_1+\beta_1-2}{\beta_1(\beta_0\beta_1-1)}\right]-1 = \frac{\beta_1-1}{\beta_0\beta_1-1} \, ,$$

$$x_3 = \beta_0[x_2]-1 = \beta_0\left[\frac{\beta_1-1}{\beta_0\beta_1-1}\right]-1 = \frac{1-\beta_0}{\beta_0\beta_1-1} \, ,$$

$$x_4 = \beta_1[x_3]+1 = \beta_1\left[\frac{1-\beta_0}{\beta_0\beta_1-1}\right]+1 = \frac{\beta_1-1}{\beta_0\beta_1-1} = x_2 \, ,$$

gives us the following period-2 pattern with two transient terms:

$$\left[\frac{\beta_0\beta_1[\beta_1+1]-2}{\beta_0\beta_1(\beta_0\beta_1-1)} \, , \, \frac{\beta_1[\beta_0+1]-2}{\beta_1(\beta_0\beta_1-1)}\right] \, , \, \frac{\beta_1-1}{\beta_0\beta_1-1} \, , \, \frac{1-\beta_0}{\beta_0\beta_1-1} \, , \, \dots \, .$$

The following graph is an example of (Figure 3.23):

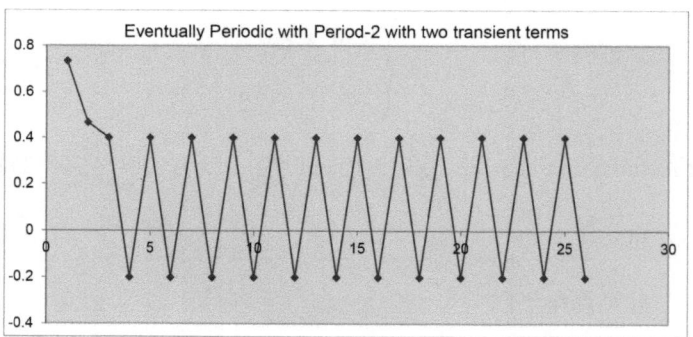

Fig. 3.23 an eventually periodic period-2 cycle with $\beta_0 = 2$, $\beta_1 = 3$, and $x_0 = \frac{11}{15}$; two transient terms of the pattern are positive and are decreasing every other term in two subsequences.

Example 27. Determine the period-2 solution with four transient terms of:

$$x_{n+1} = \beta_n x_n - g(x_n) \, , \quad n = 0,1,2, \dots \, ,$$

where $x_0 \in \mathfrak{R}$, $\{\beta_n\}_{n=0}^{\infty}$ is a period-2 sequence, and

$$g(x) = \begin{cases} 1 & \text{if } x \geq 0 \, , \\ -1 & \text{if } x < 0 \, . \end{cases}$$

Solution: Assume that $x_6 = x_4$, $\beta_0\beta_1 > 1$, and $\frac{[\beta_1+1][\beta_0\beta_1+1]}{[\beta_0\beta_1]^2} < x_0 < \frac{\beta_0[\beta_1+1][\beta_0\beta_1+1]+1}{\beta_0[\beta_0\beta_1]^2}$.
Then:

$$x_0 > 0\,,$$

$$x_1 = \beta_0 x_0 - 1\,,$$

$$x_2 = \beta_1[x_1] - 1 = \beta_1[\beta_0 x_0 - 1] - 1 = \beta_0\beta_1 x_0 - \beta_1 - 1\,,$$

$$x_3 = \beta_0[x_2] - 1 = \beta_0[\beta_0\beta_1 x_0 - \beta_1 - 1] - 1\,,$$

$$= \beta_0^2\beta_1 x_0 - \beta_0\beta_1 - \beta_0 - 1\,,$$

$$x_4 = \beta_1[x_3] - 1 = \beta_1[\beta_0^2\beta_1 x_0 - \beta_0\beta_1 - \beta_0 - 1] - 1\,,$$

$$= \beta_0^2\beta_1^2 x_0 - \beta_0\beta_1^2 - \beta_0\beta_1 - \beta_1 - 1\,,$$

$$x_5 = \beta_0[x_4] - 1 = \beta_0[\beta_0^2\beta_1^2 x_0 - \beta_0\beta_1^2 - \beta_0\beta_1 - \beta_1 - 1] - 1\,,$$

$$= \beta_0^3\beta_1^2 x_0 - \beta_0^2\beta_1^2 - \beta_0^2\beta_1 - \beta_0\beta_1 - \beta_0 - 1\,,$$

$$x_6 = \beta_1[x_4] + 1 = \beta_1[\beta_0^3\beta_1^2 x_0 - \beta_0^2\beta_1^2 - \beta_0^2\beta_1 - \beta_0\beta_1 - \beta_0 - 1] + 1\,,$$

$$= \beta_0^3\beta_1^3 x_0 - \beta_0^2\beta_1^3 - \beta_0^2\beta_1^2 - \beta_0\beta_1^2 - \beta_0\beta_1 - \beta_1 + 1 = x_4\,.$$

We see that $x_0 = \frac{(\beta_0\beta_1)^2[\beta_1+1]-2}{(\beta_0\beta_1)^2(\beta_0\beta_1-1)}$, where $\beta_0\beta_1 > 1$ gives us the following period-2 pattern with four transient terms:

$$\left[\frac{(\beta_0\beta_1)^2[\beta_1+1]-2}{(\beta_0\beta_1)^2(\beta_0\beta_1-1)}\,,\ \frac{\beta_0\beta_1^2[\beta_0+1]-2}{\beta_0\beta_1^2(\beta_0\beta_1-1)}\,,\ \frac{\beta_0\beta_1[\beta_1+1]-2}{\beta_0\beta_1(\beta_0\beta_1-1)}\,,\ \frac{\beta_1[\beta_0+1]-2}{\beta_1(\beta_0\beta_1-1)}\right]\,.$$

The graph below is an example of (Figure 3.24):

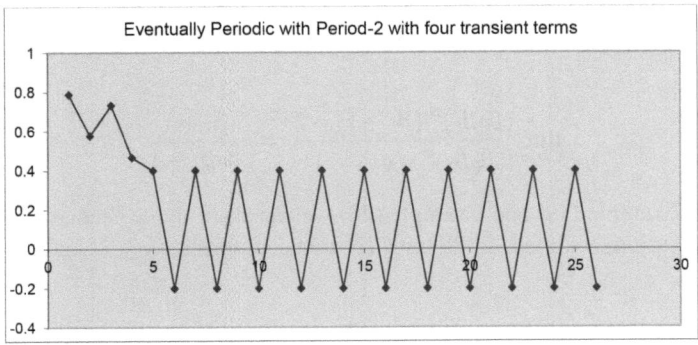

Fig. 3.24 an eventually periodic cycle with period-2 where $\beta_0 = 2$, $\beta_1 = 3$ and $x_0 = \frac{71}{90}$ and $x_0 = \frac{11}{15}$. In addition, four transient terms of the pattern are positive and are decreasing every other term in two subsequences as we observed in Examples 19 and 20.

The following graph exhibits (Figure 3.25):

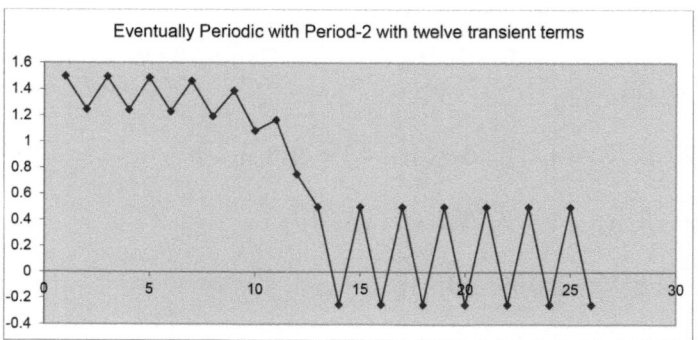

Fig. 3.25 an eventually periodic cycle with period-2; $\beta_0 = \frac{3}{2}$, $\beta_1 = 2$, and $x_0 = \frac{2185}{1458}$. In addition, twelve transient terms of the pattern are positive and are decreasing every other term in two subsequences.

From Example 19 and Example 20 we can inductively determine the pattern of the 2N transient terms, $(N \geq 1)$. The following theorem generalizes the results and the proof will be left as an exercise at the end of the chapter.

Theorem 9. *The $\Delta.E.$:*

$$x_{n+1} = \beta_n x_n - g(x_n) \ , \quad n = 0, 1, 2, \ldots ,$$

where $x_0 \in \mathfrak{R}$, $\{\beta_n\}_{n=0}^{\infty}$ is a period-2 sequence, $\beta_0 \beta_1 \neq 1$, and

$$g(x) = \begin{cases} 1 & \text{if } x \geq 0 , \\ -1 & \text{if } x < 0 , \end{cases}$$

is eventually periodic with period-2 with 2N-transient terms, $(N \in \mathbb{N})$ when:

$$x_0 = \frac{(\beta_0 \beta_1)^N (\beta_1 + 1) - 2}{(\beta_0 \beta_1)^N (\beta_0 \beta_1 - 1)} .$$

Observe that

$$\lim_{N \to \infty} \frac{(\beta_0 \beta_1)^N (\beta_1 + 1) - 2}{(\beta_0 \beta_1)^N (\beta_0 \beta_1 - 1)} = \frac{\beta_1 + 1}{\beta_0 \beta_1 - 1} .$$

Similar to Example (19) and Example (20) we can show the existence of eventually periodic solutions with period-2k with 2N-transient terms; $(k \geq 2)$ and $(N \in \mathbb{N})$.

3.10 Nonautonomous Piece-wise Difference Equation as a Neuron Model When $\beta_0\beta_1 = 1$

In this section we will study the periodic solutions and eventually periodic solutions of the Nonautonomous Piece-wise Δ.E.:

$$x_{n+1} = \beta_n x_n - g(x_n) \ , \ n = 0, 1, 2, \ldots , \tag{3.20}$$

where $x_0 \in \mathfrak{R}$, $\{\beta_n\}_{n=0}^{\infty}$ is a period-2 sequence, $\beta_0\beta_1 = 1$, and

$$g(x) = \begin{cases} 1 & \text{if } x \geq 0 , \\ -1 & \text{if } x < 0 . \end{cases}$$

We will show that Equation (3.20) only has period-4 cycles when $\beta_0\beta 1 = 1$. In fact, we will see major contrasts compared to the results with the previous section when $\beta_0\beta_1 \neq 1$. The primary difference is that we will not have unique periodic cycles.

Example 28. Determine a period-4 cycle of:

$$x_{n+1} = \beta_n x_n - g(x_n) \ , \ n = 0, 1, 2, \ldots ,$$

where $x_0 \in \mathfrak{R}$, $\{\beta_n\}_{n=0}^{\infty}$ is a period-2 sequence, $\beta_0\beta_1 = 1$, and

$$g(x) = \begin{cases} 1 & \text{if } x \geq 0 , \\ -1 & \text{if } x < 0 . \end{cases}$$

Solution: Since $\beta_0\beta_1 = 1$, then either $\beta_0 > 1$ and $\beta_1 < 1$ or vice versa, which leads to two cases:

- Case 1: Suppose that $\beta_0 > 1$, $\beta_1 < 1$, and $x_0 = -1$. Then we obtain the following period-4 pattern:
$$-1 \ , \ 1 - \beta_0 \ , \ \beta_1 \ , \ 0 \ , \ldots .$$
- Case 2: Suppose that $\beta_0 < 1$, $\beta_1 > 1$, and $x_0 = 0$. Then we procure the following period-4 pattern:
$$0 \ , \ -1 \ , \ 1 - \beta_1 \ , \ \beta_0 \ , \ldots .$$

From Case 1 and Case 2, we can see difference patterns. We can also show that Equation (3.20) has more period-4 cycles as the period-4 cycles will not be unique. This will be left as an exercise at the end of the chapter. In addition, Equation (3.20) has eventually period-4 cycles portrayed in the next graph (Figure 3.26).

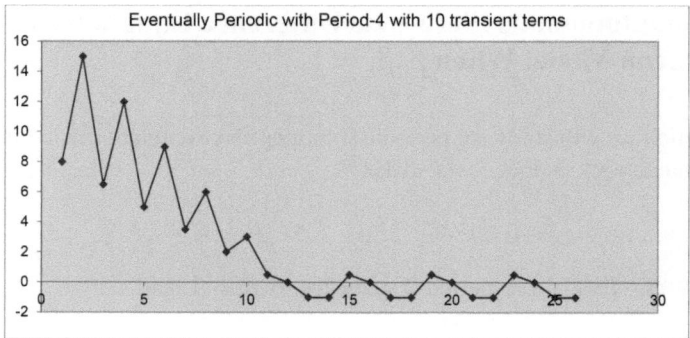

Fig. 3.26 Eventually periodic cycle with period-4; $\beta_0 = 2$, $\beta_1 = \frac{1}{2}$, and $x_0 = 8$. In addition, ten transient terms of the pattern are positive and are decreasing every other term in two linear subsequences.

On one hand, we can see some similarity to the autonomous case when $\beta = 1$. On the other hand, we will see many differences as there are many more cases and subintervals that we will have to decompose into. In the next example we will consider the case when $\beta_0\beta_1 = 1$, then $\beta_0 < 1$ and $\beta_1 > 1$.

Example 29. Determine the period-4 solution with transient terms of:

$$x_{n+1} = \beta_n x_n - g(x_n) \ , \quad n = 0, 1, 2, \ \ldots \ ,$$

where $x_0 = \beta_1 + 2$, $\{\beta_n\}_{n=0}^{\infty}$ is a period-2 sequence, $\beta_0\beta_1 = 1$, and

$$g(x) = \begin{cases} 1 & \text{if } x \geq 0 \, , \\ -1 & \text{if } x < 0 \, . \end{cases}$$

Solution: Suppose:

(1) $\beta_0\beta_1 = 1$,

(2) $\beta_0 < 1$ and $\beta_1 > 1$

(3) $\frac{3}{2} < \beta_1 < 2$.

Then we acquire:

$$x_0 = \beta_1 + 2 \, ,$$

$$x_1 = \beta_0 [x_0] - 1 = \beta_0\beta_1 + 2\beta_0 - 1 = 2\beta_0 \, ,$$

$$x_2 = \beta_1 [x_1] - 1 = 2\beta_0\beta_1 - 1 = 1 \, ,$$

$$x_3 = \beta_0 [x_2] - 1 = \beta_0 - 1 < 0 \,,$$

$$x_4 = \beta_1 [x_3] + 1 = \beta_0\beta_1 - \beta_1 + 1 = 2 - \beta_1 > 0 \,,$$

$$x_5 = \beta_0 [x_4] - 1 = 2\beta_0 - \beta_0\beta_1 - 1 = 2\beta_0 - 2 = 2(\beta_0 - 1) < 0 \,,$$

$$x_6 = \beta_1 [x_5] + 1 = 2\beta_0\beta_1 - 2\beta_1 + 1 = 3 - 2\beta_1 < 0 \,,$$

$$x_7 = \beta_0 [x_6] + 1 = 3\beta_0 - 2\beta_0\beta_1 + 1 = 3\beta_0 - 1 > 0 \,,$$

$$x_8 = \beta_1 [x_7] + 1 = 3\beta_0\beta_1 - \beta_1 - 1 = 2 - \beta_1 = x_4 \,.$$

Therefore, there are four transient terms and we conclude that there exists $k \in \mathbb{N}$ such that:

$$\frac{k+2}{k+1} < \beta_1 < \frac{k+1}{k} \,.$$

as we can see in the diagram below (Figure 3.27).

$$\frac{7}{6} \qquad \frac{6}{5} \qquad \frac{5}{4} \qquad \frac{4}{3} \qquad \frac{3}{2} \qquad 2$$

Fig. 3.27 We choose $\beta_1 \in \left[\frac{k+2}{k+1}, \frac{k+1}{k}\right]$ for some $k \in \mathbb{N}$.

From Example 29 we pose the following theorem.

Theorem 10. *The $\Delta.E.$:*

$$x_{n+1} = \beta_n x_n - g(x_n) \,, \quad n = 0, 1, 2, \ldots \,,$$

where $x_0 \in \mathfrak{R}$, $\{\beta_n\}_{n=0}^{\infty}$ is a period-2 sequence, $\beta_0\beta_1 = 1$, and

$$g(x) = \begin{cases} 1 & \text{if } x \geq 0 \,, \\ -1 & \text{if } x < 0 \,, \end{cases}$$

is eventually periodic with period-4 if:

(1) $\beta_0 < 1$ and $\beta_1 > 1$

(2) $x_0 = \beta_1 + 2$

(3) For some $k \in \mathbb{N}$:

$$\frac{k+2}{k+1} < \beta_1 < \frac{k+1}{k} \,.$$

The proof of Theorem 10 will be left as an exercise at the end of the chapter. Notice that if $\beta_1 = \frac{k+2}{k+1}$ ($k \in \mathbb{N}$), then the transient terms will merge to the following period-4 cycle:

$$0 \, , \, -1 \, , \, 1 - \beta_1 \, , \, \beta_0 \, , \, \dots \, .$$

Observe that this period-4 cycle appears in Case 2 of Example 28. Moreover, applying the initial condition $x_0 = \beta + N$ for $N \in \mathbb{N}$ can also work to find eventually periodic solutions with period-4. There are many cases to consider to prove the result that is still research work in progress.

More studies and inquiries on piece-wise difference equations will be addressed in Chapter 6. In addition, the following questions will be conveyed regarding the periodic character of solutions:

- When $\{\beta_n\}_{n=0}^{\infty}$ is a period-k sequence, $(k \geq 3)$, will periodic and eventually periodic solutions exist with only period-3N, $(N \geq 1)$?

- Systems of piece-wise difference equations.

3.11 Exercises

1.   Consider the **Nonautonomous Riccati** Δ.**E.** in the form:

$$x_{n+1} = \frac{1}{x_n} + a_n \, , \quad n = 0, 1, \dots \, ,$$

where $\{a_n\}_{n=0}^{\infty}$ is a period-2 sequence. Show that there are two unique period-2 cycles and determine their patterns.

Consider the **Nonautonomous Riccati** Δ.**E.** in the form:

$$x_{n+1} = \frac{a_n}{x_n} \, , \quad n = 0, 1, \dots \, .$$

where $\{a_n\}_{n=0}^{\infty}$ is a periodic sequence. In problems 2–9:

2.   Show that there is no period-3 solution when $\{a_n\}_{n=0}^{\infty}$ is a period-2 sequence and explain why.

3.   Show that there is no period-4 solution when $\{a_n\}_{n=0}^{\infty}$ is a period-2 sequence and explain why.

4.   Show that there is no period-5 solution when $\{a_n\}_{n=0}^{\infty}$ is a period-2 sequence and explain why.

5.   From Exercises 2, 3, and 4, show that there is no period-k $(k \geq 2)$ solution when $\{a_n\}_{n=0}^{\infty}$ is a period-2 sequence and explain why.

6.   Suppose that $\{a_n\}_{n=0}^{\infty}$ is a period-5 sequence. Determine the pattern of the periodic cycle.

7.   Suppose that $\{a_n\}_{n=0}^{\infty}$ is a period-6 sequence. Determine the pattern of the periodic cycle.

8. Suppose that $\{a_n\}_{n=0}^{\infty}$ is a period-$(2k+1)$ sequence (for $k \in \mathbb{N}$). From Exercise 5, determine the pattern of the periodic cycle.

9. Suppose that $\{a_n\}_{n=0}^{\infty}$ is a period-$(2k)$ sequence for (for $k \geq 2$). From Exercise 6, determine the pattern of the periodic cycle.

Consider the **Nonautonomous Riccati Δ.E.** in the form:

$$x_{n+1} = \frac{a_n x_n}{x_n - 1} \quad , \quad n = 0, 1, \ldots .$$

where $\{a_n\}_{n=0}^{\infty}$ is a periodic sequence. In problems 10–15:

10. Suppose that $\{a_n\}_{n=0}^{\infty}$ is a period-6 sequence. Determine the pattern of the periodic cycle.

11. Suppose that $\{a_n\}_{n=0}^{\infty}$ is a period-8 sequence. Determine the pattern of the periodic cycle.

12. Suppose that $\{a_n\}_{n=0}^{\infty}$ is a period-2k sequence for (for $k \geq 2$). From Exercises 10 and 11, determine the pattern of the periodic cycle.

13. Suppose that $\{a_n\}_{n=0}^{\infty}$ is a period-3 sequence. Determine the pattern of the periodic cycle. **Hint:** set $x_6 = x_0$ and proceed with the algebra.

14. Suppose that $\{a_n\}_{n=0}^{\infty}$ is a period-5 sequence. Determine the pattern of the periodic cycle. **Hint:** set $x_{10} = x_0$ and proceed with the algebra.

15. Suppose that $\{a_n\}_{n=0}^{\infty}$ is a period-$(2k+1)$, $(k \in \mathbb{N})$. From Exercises 13 and 14, determine the pattern of the periodic cycle.

Consider the **Logistic Δ.E.**:

$$x_{n+1} = 4x_n(1 - x_n) \quad , \quad n = 0, 1, \ldots .$$

In problems 16–22, using the double angle identity $sin(2t) = 2sin(t)cos(t)$:

16. Determine the periodic pattern when $x_0 = sin^2\left(\frac{2\pi}{3}\right)$.

17. Determine the periodic pattern when $x_0 = sin^2\left(\frac{2\pi}{9}\right)$.

18. Determine the periodic pattern when $x_0 = sin^2\left(\frac{2\pi}{11}\right)$.

19. Determine the periodic pattern when $x_0 = sin^2\left(\frac{2\pi}{13}\right)$.

20. Determine the periodic pattern when $x_0 = sin^2\left(\frac{\pi}{12}\right)$.

21. Determine the periodic pattern when $x_0 = sin^2\left(\frac{\pi}{96}\right)$.

22. Determine the periodic pattern when $x_0 = sin^2\left(\frac{\pi}{160}\right)$.

Consider the **Tent-Map** in the form:

$$x_{n+1} = \begin{cases} 2x_n & \text{if } x_n < \frac{1}{2}, \\ 2(1-x_n) & \text{if } x_n \geq \frac{1}{2}. \end{cases} \qquad n = 0, 1, \dots,$$

where $0 < x_0 < 1$. In problems 23–28:

23. Determine a period-5 pattern.

24. Determine a period-6 pattern.

25. From Exercises 23 and 24, determine a period-k pattern (for $k \geq 2$).

26. Determine a period-6 pattern with four transient terms.

27. Determine a period-12 pattern with eight transient terms.

28. From Exercises 26 and 27, determine a period-k pattern (for $k \geq 2$) with N transient terms ($N \in \mathbb{N}$).

Consider the **Tent-Map** in the form:

$$x_{n+1} = \begin{cases} Ax_n & \text{if } x_n < \frac{1}{2}, \\ A(1-x_n) & \text{if } x_n \geq \frac{1}{2}. \end{cases} \qquad n = 0, 1, \dots,$$

where $0 < x_0 < 1$ and $A > 2$. In problems 29–35:

29. Determine a period-2 pattern.

30. Determine a period-3 pattern.

31. Determine a period-4 pattern.

32. From Exercises 29, 30, and 31, determine a period-k pattern (for $k \geq 2$).

33. Determine a period-2 pattern with four transient terms.

34. Determine a period-4 pattern with eight transient terms.

35. From Exercises 33 and 34, determine a period-k pattern (for $k \geq 2$) with N transient terms ($N \in \mathbb{N}$).

Consider the **Piece-wise Δ.E.** in the form:

$$x_{n+1} = \beta x_n - g(x_n), \quad n = 0, 1, 2, \dots,$$

where $\beta > 0$ and

$$g(x) = \begin{cases} 1, & x \geq 0, \\ -1, & x < 0. \end{cases}$$

In problems 36–47:

36. Determine a period-4 pattern.

37. Determine a period-5 pattern.

38. From Exercise 36, determine a pattern of the period-2k cycle, $(k \in \mathbb{N})$.

39. From Exercise 37, determine a pattern of the period-$(2k+1)$ cycle, $(k \in \mathbb{N})$.

40. Determine a period-2 pattern with 6 transient terms.

41. Determine a period-2 pattern with 9 transient terms.

42. From Exercises 40 and 41, determine a period-2 pattern with N transient terms, $(N \in \mathbb{N})$.

43. Determine a period-3 pattern with 2 transient terms.

44. Determine a period-3 pattern with 3 transient terms.

45. From Exercises 42 and 43, determine a period-3 pattern with N transient terms, $(N \in \mathbb{N})$.

46. From Exercise 42, determine a period-2k pattern, $(k \in \mathbb{N})$ with N transient terms, $(N \in \mathbb{N})$.

47. From Exercise 45, determine a period-$(2k+1)$ pattern, $(k \in \mathbb{N})$ with N transient terms, $(N \in \mathbb{N})$.

Consider the **Nonautonomous Piece-wise** Δ.**E.** in the form:

$$x_{n+1} = \beta_n x_n - g(x_n) \ , \quad n = 0, 1, 2, \ldots ,$$

where $\{\beta_n\}_{n=0}^{\infty}$ is a period-2 sequence and

$$g(x) = \begin{cases} 1, & x \geq 0, \\ -1, & x < 0. \end{cases}$$

In problems 48–65:

48. Determine a period-6 pattern.

49. Determine a period-8 pattern.

50. From Exercises 48 and 49, determine a period-2k pattern, $(k \in \mathbb{N})$.

51. Determine a period-2 pattern with 6 transient terms.

52. Determine a period-2 pattern with 8 transient terms.

53. From Exercises 51 and 52, determine a period-2 pattern with 2N transient terms, $(N \in \mathbb{N})$.

54. Determine a period-4 pattern with 2 transient terms.

55. Determine a period-4 pattern with 4 transient terms.

56. From Exercises 54 and 55, determine a period-4 pattern with 2N transient terms, $(N \in \mathbb{N})$.

57. From Exercises 53 and 56, determine a period-2k pattern, $(k \in \mathbb{N})$ with 2N transient terms, $(N \in \mathbb{N})$.

58. Suppose that $\beta_0\beta_1 = 1$, $\beta_0 < 1$, $\frac{4}{3} < \beta_1 < \frac{3}{2}$ and let $x_0 = \beta_1 + 2$. Determine the period-4 solution and the pattern of the transient terms.

59. Suppose that $\beta_0\beta_1 = 1$, $\beta_0 < 1$, $\frac{5}{4} < \beta_1 < \frac{4}{3}$ and let $x_0 = \beta_1 + 2$. Determine the period-4 solution and the pattern of the transient terms.

60. Using Exercises 52 and 53, suppose that $\beta_0\beta_1 = 1$, $\beta_0 < 1$, $\frac{k+2}{k+1} < \beta_1 < \frac{k+1}{k}$ (for $k \in \mathbb{N}$) and let $x_0 = \beta_1 + 2$. Determine the period-4 solution and the pattern of the transient terms.

61. Suppose that $\beta_0\beta_1 = 1$, $\beta_0 < 1$, $\beta_1 = \frac{4}{3}$ and let $x_0 = \beta_1 + 2$. Determine the period-4 solution and the pattern of the transient terms. Compare similarities and differences with Exercise 52.

62. Suppose that $\beta_0\beta_1 = 1$, $\beta_0 < 1$, $\beta_1 = \frac{6}{5}$ and let $x_0 = \beta_1 + 2$. Determine the period-4 solution and the pattern of the transient terms. Compare similarities and differences with Exercise 52.

63. Using Exercises 53 and 54, suppose that $\beta_0\beta_1 = 1$, $\beta_0 < 1$, $\beta_1 = \frac{k+2}{k+1}$ (for $k \geq 1$) and let $x_0 = \beta_1 + 2$. Determine the period-4 solution and the pattern of the transient terms. Compare similarities and differences with Exercise 53.

64. Suppose that $\beta_0\beta_1 = 1$, $\beta_0 = \frac{1}{k}$, $\beta_1 = k$ and let $x_0 = \beta_1 + k$ $(k > 1)$. Determine the period-4 solution and the pattern of the transient terms.

65. Suppose that $\beta_0\beta_1 = 1$, $\beta_0 = \frac{1}{k-1}$, $\beta_1 = k - 1$ and let $x_0 = \beta_1 + k$ $(k > 2)$. Determine the period-4 solution and the pattern of the transient terms.

Chapter 4
Second Order Linear Difference Equations and Periodic Traits

4.1 Second Order Linear Difference Equations

This chapter's aims are to solve second order linear difference equations explicitly, to determine the necessary and sufficient conditions for the existence and uniqueness of periodic solutions, and to determine specific patterns of periodic cycles of second order linear difference equations. We will break up this chapter into three sections: Periodic Solutions of Second Order Linear Homogeneous Difference Equations, Second Order Nonhomogeneous Difference Equations, and Second Order Nonautonomous Linear Difference Equations. We will commence with three examples of second order linear difference equations:

(i)   (Homogeneous) $x_{n+2} = 3x_{n+1} - x_n$, $n = 0, 1, \ldots$.

(ii)   (Nonhomogeneous) $x_{n+2} = 4x_n + 3$, $n = 0, 1, \ldots$.

(iii)   (Nonautonomous) $x_{n+2} = x_{n+1} + 4x_n - \left(\frac{2}{3}\right)^n$, $n = 0, 1, \ldots$.

Our objective is to solve each Δ.E. explicitly, check the solution, analyze the long-term behavior of solutions, and analyze the patterns of the periodic cycles. We will start our study with the **Second order Linear Homogeneous Δ.E.** in the form:

$$x_{n+2} + px_{n+1} + qx_n = 0 \ , \quad n = 0, 1, \ldots ,$$

where $p \in \mathfrak{R}$, $q \neq 0$, and the initial conditions $x_0, x_1 \in \mathfrak{R}$.

4.2 Second Order Homogeneous Linear Difference Equations

The following are examples of second order linear homogeneous difference equations:

(i) $x_{n+2} = x_{n+1} + x_n$, $n = 0, 1, \ldots$.

(ii) $4x_{n+2} = x_n$, $n = 0, 1, \ldots$.

© Springer Nature Switzerland AG 2018
M. A. Radin, *Periodic Character and Patterns of Recursive Sequences*,
https://doi.org/10.1007/978-3-030-01780-4_4

We will proceed with the second order linear Δ.E. in the form:

$$x_{n+2} + px_{n+1} + qx_n = 0 \ , \quad n = 0, 1, \ldots , \tag{4.1}$$

where $p \in \mathfrak{R}$, $q \neq 0$, and the initial conditions $x_0, x_1 \in \mathfrak{R}$. To obtain an explicit solution, we will set $x_n = \lambda^n$, which produces the following quadratic **characteristic polynomial**:

$$\lambda^2 + p\lambda + q = 0 , \tag{4.2}$$

whose roots (solutions) are

$$\lambda_1 = \frac{-p + \sqrt{p^2 - 4q}}{2} \quad \text{and} \quad \lambda_2 = \frac{-p - \sqrt{p^2 - 4q}}{2} .$$

Also let $D = p^2 - 4q$; either $D > 0$, $D = 0$ or $D < 0$, which leads to the following three cases:

CASE 1: Suppose that $D > 0$. Then Equation (4.2) has two distinct real roots $\lambda_1 \neq \lambda_2$. Then we obtain the following solution to Equation (4.1):

$$x_n = C_1(\lambda_1^n) + C_2(\lambda_2^n) \ , \quad n = 0, 1, \ldots . \tag{4.3}$$

CASE 2: Suppose that $D = 0$. Then Equation (4.2) has a repeated real root $\lambda_1 = \lambda_2$. Then we acquire the following solution to Equation (4.1):

$$x_n = C_1(\lambda_1^n) + C_2 n(\lambda_1^n) \ , \quad n = 0, 1, \ldots . \tag{4.4}$$

CASE 3: Suppose that $D < 0$. Then Equation (4.2) has two imaginary complex conjugate roots

$$\lambda_1 = A + Bi \quad \text{and} \quad \lambda_2 = A - Bi .$$

Now let

$$\Lambda = |A + Bi| = |A - Bi| \quad \text{and} \quad \theta = arg(\lambda_1).$$

Then we procure the following solution to Equation (4.1):

$$x_n = (\Lambda)^n [C_1 sin(n\theta) + C_2 cos(n\theta)] \ , \quad n = 0, 1, \ldots . \tag{4.5}$$

The following examples will demonstrate the implementation of the three cases above.

Example 1. Solve:

$$x_{n+2} - 5x_{n+1} + 6x_n = 0 \ , \quad n = 0, 1, \ldots , \tag{4.6}$$

and verify that the solution is correct.

Solution: The **characteristic polynomial** of Equation (4.6) is:

$$\lambda^2 - 5\lambda + 6 = 0 ,$$

whose roots are $\lambda_1 = 2$ and $\lambda_2 = 3$. So via (4.3):

$$x_n = C_1 2^n + C_2 3^n \ , \quad n = 0, 1, \dots . \tag{4.7}$$

To check our solution, via (4.7) we get:

$$x_{n+1} = C_1 2^{n+1} + C_2 3^{n+1} = 2C_1 2^n + 3C_2 3^n , \tag{4.8}$$

and

$$x_{n+2} = C_1 2^{n+2} + c_2 3^{n+2} = 4C_1 2^n + 9C_2 3^n . \tag{4.9}$$

By substituting (4.7), (4.8), and (4.9) into Equation (4.6) we get:

$$x_{n+2} - 5x_{n+1} + 6x_n$$

$$= [4C_1 2^n + 9C_2 3^n] - 5[2C_1 2^n + 3C_2 3^n] + 6[C_1 2^n + C_2 3^n]$$

$$= 4C_1 2^n + 9C_2 3^n - 10C_1 2^n - 15C_2 3^n + 6C_1 2^n + 6C_2 3^n = 0.$$

Hence the result follows.

Example 2. Solve:

$$x_{n+2} - 6x_{n+1} + 9x_n = 0 \ , \quad n = 0, 1, \dots , \tag{4.10}$$

and verify that the solution is correct.

Solution: The **characteristic polynomial** of Equation (4.10) is:

$$\lambda^2 - 6\lambda + 9 = 0 ,$$

and has a repeated root $\lambda_1 = \lambda_2 = 3$. So via (4.4):

$$x_n = C_1 3^n + C_2 n 3^n \ , \quad n = 0, 1, \dots . \tag{4.11}$$

To check our solution, via (4.11) we acquire:

$$x_{n+1} = C_1 3^{n+1} + C_2 (n+1) 3^{n+1} = 3C_1 3^n + 3C_2 (n+1) 3^n , \tag{4.12}$$

and

$$x_{n+2} = C_1 3^{n+2} + C_2 (n+2) 3^{n+2} = 9C_1 2^n + 9C_2 (n+2) 3^n . \tag{4.13}$$

By substituting (4.11), (4.12), and (4.13) into Equation (4.10) we obtain:

$$x_{n+2} - 6x_{n+1} + 9x_n$$

$$= [9C_1 3^n + 9C_2 (n+2) 3^n] - 6[3C_1 3^n + 3C_2 (n+1) 3^n] + 9[C_1 3^n + C_2 n 3^n]$$

$$= 0 .$$

The result follows.

Example 3. Solve:

$$x_{n+2} + x_n = 0 \ , \ n = 0, 1, \dots , \tag{4.14}$$

and verify that the solution is correct.

Solution: The **characteristic polynomial** of Equation (4.14) is:

$$\lambda^2 + 1 = 0 ,$$

and has two imaginary roots $\lambda_1 = i$ and $\lambda_2 = -i$. Also note that

$$\Lambda = |i| = |-i| = 1 \ \text{ and } \ \theta = arg(i) = \frac{\pi}{2} .$$

Therefore via (4.5):

$$x_n = C_1 sin\left(\frac{n\pi}{2}\right) + C_2 cos\left(\frac{n\pi}{2}\right) \ , \ n = 0, 1, \dots . \tag{4.15}$$

To check our solution, via (4.15) we acquire:

$$x_{n+2} = C_1 sin\left(\frac{[n+2]\pi}{2}\right) + C_2 cos\left(\frac{[n+2]\pi}{2}\right) . \tag{4.16}$$

Now by substituting (4.15) and (4.16) into Equation (4.14) we acquire:

$$x_{n+2} + x_n$$

$$= \left[C_1 sin\left(\frac{[n+2]\pi}{2}\right) + C_2 cos\left(\frac{[n+2]\pi}{2}\right)\right] + \left[C_1 sin\left(\frac{n\pi}{2}\right) + C_2 cos\left(\frac{n\pi}{2}\right)\right]$$

$$= C_1 \left[sin\left(\frac{n\pi}{2}\right) cos(\pi) + sin(\pi) cos\left(\frac{n\pi}{2}\right)\right] + C_2 \left[cos\left(\frac{n\pi}{2}\right) cos(\pi) - sin\left(\frac{n\pi}{2}\right) sin(\pi)\right]$$

$$+ \left[C_1 sin\left(\frac{n\pi}{2}\right) + C_2 cos\left(\frac{n\pi}{2}\right)\right]$$

$$= -C_1 sin\left(\frac{n\pi}{2}\right) - C_2 cos\left(\frac{n\pi}{2}\right) + C_1 sin\left(\frac{n\pi}{2}\right) + C_2 cos\left(\frac{n\pi}{2}\right) = 0 .$$

Hence the result follows. Furthermore, we procure period-4 cycles as $\theta = \frac{\pi}{2}$.

Example 4. Solve the Initial Value Problem:

$$\begin{cases} x_{n+2} - 3x_{n+1} + 2x_n = 0 \ , \ n = 0, 1, \dots , \\ \\ x_0 = 2 , \\ \\ x_1 = 3 , \end{cases}$$

and verify that the solution is correct.

Solution: Via (4.3):

$$x_n = C_1 1^n + C_2 2^n = C_1 + C_2 2^n \ , \quad n = 0, 1, \dots \ . \tag{4.17}$$

First we solve for C_1 and C_2. Via (4.17) we produce:

$$x_0 = C_1 + C_2 = 2 \, ,$$

$$x_1 = C_1 + 2C_2 = 3 \, .$$

Notice $C_1 = 2 - C_2$ and:

$$x_1 = C_1 + 2C_2 = (2 - C_2) + 2C_2 = 2 + C_2 = 3 \, .$$

Hence:

$$C_2 = 1 \ \text{ and } \ C_1 = 1 \, .$$

Therefore, via (4.17) the specific solution is:

$$x_n = C_1 1^n + C_2 2^n = 1 + 2^n \ , \quad n = 0, 1, \dots \ . \tag{4.18}$$

Also observe via (4.18) we produce:

$$x_{n+1} = 1 + 2^{n+1} = 1 + 2(2^n) \, , \tag{4.19}$$

and

$$x_{n+2} = 1 + 2^{n+2} = 1 + 4(2^n) \, . \tag{4.20}$$

By substituting (4.18), (4.19), and (4.20) into the given Δ.E., we get:

$$x_{n+2} - 3x_{n+1} + 2x_n$$

$$= [1 + 4(2^n)] - 3[1 + 2(2^n)] + 2[1 + 2^n]$$

$$= 1 + 4(2^n) - 3 - 6(2^n) + 2 + 2^n = 0 \, .$$

The result follows.

Example 5. Solve the Initial Value Problem:

$$\begin{cases} x_{n+2} - 4x_{n+1} + 4x_n = 0 \ , \quad n = 0, 1, \dots \, , \\[2mm] x_0 = 1 \, , \\[2mm] x_1 = 6 \, , \end{cases}$$

and verify that the solution is correct.

Solution: The general solution is

$$x_n = C_1 2^n + C_2 n 2^n \ , \quad n = 0, 1, \dots \ . \tag{4.21}$$

To solve for C_1 and C_2 set:

$$x_0 = C_1 = 1 \ ,$$

$$x_1 = 2C_1 + 2C_2 = 6 \ ,$$

which gives us:

$$C_1 = 1 \quad \text{and} \quad C_2 = 2 \ .$$

Therefore, via (4.21) the specific solution is:

$$x_n = C_1 2^n + C_2 n 2^n = 2^n + 2n 2^n \ , \quad n = 0, 1, \dots \ . \tag{4.22}$$

Furthermore:

$$x_{n+1} = 2^{n+1} + 2(n+1)2^{n+1} = 2(2^n) + 4(n+1)2^n, \tag{4.23}$$

and

$$x_{n+2} = 2^{n+2} + 2(n+2)2^{n+2} = 4(2^n) + 8(n+2)2^n. \tag{4.24}$$

By substituting (4.22), (4.23), and (4.24) into the given Δ.E., we acquire:

$$x_{n+2} - 4x_{n+1} + 4x_n$$

$$= [4(2^n) + 8(n+2)2^n] - 4[2(2^n) + 4(n+1)2^n] + 4[2^n + 2n2^n]$$

$$= 0 \ .$$

The result follows.

Example 6. We define the **Fibonacci Sequence** as:

$$\{x_n\}_{n=0}^{\infty} = 1, 1, 2, 3, 5, 8, 13, 21, \dots$$

Our aim is to show that

$$\lim_{n \to \infty} \frac{x_{n+1}}{x_n} = \phi = \frac{1 + \sqrt{5}}{2} \ .$$

First set:

$$x_0 = 1 \ ,$$

$$x_1 = 1 \ .$$

Then:

$$x_2 = 1 + 1 = x_1 + x_0 = 2,$$

$$x_3 = 2 + 1 = x_2 + x_1 = 3,$$

$$x_4 = 3 + 2 = x_3 + x_2 = 5.$$

Now we obtain the following Initial Value Problem:

$$\begin{cases} x_{n+2} = x_{n+1} + x_n, \quad n = 0, 1, \ldots, \\ \\ x_0 = 1, \\ \\ x_1 = 1, \end{cases}$$

whose **characteristic polynomial** is:

$$\lambda^2 - \lambda - 1 = 0,$$

with two distinct real roots:

$$\lambda_1 = \frac{1 + \sqrt{5}}{2} \quad \text{and} \quad \lambda_2 = \frac{1 - \sqrt{5}}{2}.$$

Thus the general solution is

$$x_n = C_1 \lambda_1^n + C_2 \lambda_2^n, \quad n = 0, 1, \ldots.$$

Now notice:

$$\lim_{n \to \infty} \frac{x_{n+1}}{x_n} = \lim_{n \to \infty} \frac{C_1 \lambda_1^{n+1} + C_2 \lambda_2^{n+1}}{C_1 \lambda_1^n + C_2 \lambda_2^n}$$

$$= \lim_{n \to \infty} \frac{C_1 \frac{\lambda_1^{n+1}}{\lambda_1^n} + C_2 \frac{\lambda_2^{n+1}}{\lambda_1^n}}{C_1 \frac{\lambda_1^n}{\lambda_1^n} + C_2 \frac{\lambda_2^n}{\lambda_1^n}} = \lambda_1 = \frac{1 + \sqrt{5}}{2} = \phi.$$

We will proceed with studying the Nonhomogeneous Second Order Linear Difference Equations in the form:

$$x_{n+2} + p x_{n+1} + q x_n = r, \quad n = 0, 1, \ldots,$$

where $x_0, x_1, p \in \Re$ and $q, r \neq 0$.

4.3 Nonhomogeneous Second Order Linear Difference Equations with a Constant Coefficient

These are examples of second order linear nonhomogeneous difference equations:

(i) $x_{n+2} = 4x_{n+1} - 3x_n + 1$, $n = 0, 1, \ldots$.

(ii) $9x_{n+2} = x_n + 2$, $n = 0, 1, \ldots$.

Our aim is to obtain an explicit solution of the second order linear nonhomogeneous Δ.E.:

$$x_{n+2} + px_{n+1} + qx_n = r , \quad n = 0, 1, \ldots , \tag{4.25}$$

where $x_0, x_1, p \in \Re$ and $q, r \neq 0$. From (4.3), (4.4), and (4.5), let

$$x_n^h , \quad n = 0, 1, \ldots ,$$

be the **homogeneous solution** of Equation (4.25). It is also our goal to determine the **particular solution**

$$x_n^P , \quad n = 0, 1, \ldots ,$$

of Equation (4.25). Let $S = 1 + p + q$. If $S \neq 0$, then Equation (4.25) has one equilibrium point $\bar{x} = \frac{r}{1+p+q}$. If $S = 0$, then Equation (4.25) has no equilibrium points. This leads the investigation to two cases:

CASE 1. Suppose that $S \neq 0$. Then $\bar{x} = \frac{r}{1+p+q}$ is the only equilibrium point of Equation (4.25) and:

$$x_n^P = \bar{x} = \frac{r}{1+p+q} , \quad n = 0, 1, \ldots . \tag{4.26}$$

Then the general solution of Equation (4.25) is:

$$x_n = x_n^h + x_n^P = x_n^h + \bar{x} , \quad n = 0, 1, \ldots . \tag{4.27}$$

CASE 2. Suppose that $S = 0$. Then Equation (4.25) has no equilibrium points and we set

$$x_n^P = An , \quad n = 0, 1, \ldots . \tag{4.28}$$

To solve for A, we substitute (4.28) into Equation (4.25) and we obtain:

$$\begin{aligned}
&x_{n+2} + px_{n+1} + qx_n \\
&= A(n+2) + pA(n+1) + qAn = An + 2A + pAn + pA + qAn \\
&= An(1 + p + q) + 2A + pA = 2A + pA = r .
\end{aligned}$$

Thus:

$$A = \frac{r}{p+2} .$$

Then the general solution of Equation (4.25) is:

$$x_n = x_n^h + x_n^P = x_n^h + An \quad , \quad n = 0, 1, \dots . \tag{4.29}$$

The following examples will illustrate the implementation of the two cases above.

Example 7. Solve:

$$x_{n+2} - 5x_{n+1} + 6x_n = 2 \quad , \quad n = 0, 1, \dots , \tag{4.30}$$

and verify that the solution is correct.

Solution: First of all, $\bar{x} = \frac{2}{1-5+6} = 1$. Second, via (4.3):

$$x_n^h = C_1 2^n + C_2 3^n \quad , \quad n = 0, 1, \dots . \tag{4.31}$$

Also, via (4.26):

$$x_n^P = \bar{x} = 1 \quad , \quad n = 0, 1, \dots . \tag{4.32}$$

Hence

$$x_n = x_n^h + x_n^P = C_1 2^n + C_2 3^n + 1 \quad , \quad n = 0, 1, \dots . \tag{4.33}$$

Then

$$x_{n+1} = x_{n+1}^h + x_{n+1}^P = C_1 2^{n+1} + C_2 3^{n+1} + 1 \quad , \quad n = 0, 1, \dots , \tag{4.34}$$

and

$$x_{n+2} = x_{n+2}^h + x_{n+2}^P = C_1 2^{n+2} + C_2 3^{n+2} + 1 \quad , \quad n = 0, 1, \dots . \tag{4.35}$$

Now we substitute (4.33), (4.34), and (4.35) into Equation (4.30) and we obtain:

$$x_{n+2} - 5x_{n+1} + 6x_n$$

$$= [4C_1 2^n + 9C_2 3^n + 1] - 5[2C_1 2^n + 3C_2 3^n + 1] + 6[C_1 2^n + C_2 3^n + 1]$$

$$= 4C_1 2^n + 9C_2 3^n + 1 - 10C_1 2^n - 15C_2 3^n - 5 + 6C_1 2^n + 6C_2 3^n + 6 = 2.$$

The result follows.

Example 8. Solve:

$$x_{n+2} - 5x_{n+1} + 4x_n = 12 \quad , \quad n = 0, 1, \dots , \tag{4.36}$$

and verify that the solution is correct.

Solution: Observe that Equation (4.36) has no equilibrium points. In addition, via (4.3):

$$x_n^h = C_1 1^n + C_2 4^n = C_1 + C_2 4^n \quad , \quad n = 0, 1, \dots . \tag{4.37}$$

To determine the Particular Solution, via (4.28), set

$$x_n^P = An \ , \quad n = 0, 1, \dots \ . \tag{4.38}$$

By substituting (4.38) into Equation (4.36) gives us:

$$x_{n+2} - 5x_{n+1} + 4x_n$$

$$= A(n+2) - 5A(n+1) + 4An$$

$$= An + 2A - 5An - 5A + 4An$$

$$= -3A = 12 \ .$$

Thus

$$A = -4 \ .$$

Hence

$$x_n^P = An = -4n \ , \quad n = 0, 1, \dots \ . \tag{4.39}$$

Therefore, the general solution of Equation (4.36) is:

$$x_n = x_n^h + x_n^P = C_1 + C_2 4^n - 4n \ , \quad n = 0, 1, \dots \ . \tag{4.40}$$

To check the general solution, it suffices to check the homogeneous and the particular solutions separately. First by substituting (4.39) into Equation (4.36) we acquire:

$$x_{n+2} - 5x_{n+1} + 4x_n$$

$$= [-4(n+2)] - 5[-4(n+1)] + 4[-4n]$$

$$= -4n - 8 + 20n + 20 - 16n = 12.$$

Now substitute (4.37) into Equation (4.36) and we procure:

$$x_{n+2} - 5x_{n+1} + 4x_n$$

$$= [C_1 + 16C_2 4^n] - 5[C_1 + 4C_2 4^n] + 4[C_1 + C_2 4^n]$$

$$= C_1 + 16C_2 4^n - 5C_1 - 20C_2 4^n + 4C_1 + 4C_2 4^n = 0 \ .$$

The result follows.

Example 9. Solve the Initial Value Problem:

$$\begin{cases} x_{n+2} - 4x_{n+1} + 3x_n = 2 \ , \quad n = 0, 1, \dots , \\ \\ x_0 = 4 \ , \\ \\ x_1 = 7 \ . \end{cases}$$

Solution: Via (4.3), the homogeneous solution of the given Δ.E. is:

$$x_n^h = C_1 1^n + C_2 3^n = C_1 + C_2 3^n \ , \quad n = 0, 1, \ldots . \tag{4.41}$$

To solve for the Particular Solution, via (4.28) we set:

$$x_n^P = An \ , \quad n = 0, 1, \ldots . \tag{4.42}$$

By substituting (4.42) into the given Δ.E. we obtain:

$$x_{n+2} - 4x_{n+1} + 3x_n$$

$$= A(n+2) - 4A(n+1) + 3An = -2A = 2 .$$

Therefore $A = -1$ and

$$x_n^P = An = -n \ , \quad n = 0, 1, \ldots . \tag{4.43}$$

The general solution to the given Δ.E. is:

$$x_n = x_n^h + x_n^P = C_1 + C_2 3^n - n \ , \quad n = 0, 1, \ldots . \tag{4.44}$$

Now we will solve for C_1 and C_2. Via (4.44):

$$x_0 = C_1 + C_2 = 4 ,$$

$$x_1 = C_1 + 3C_2 - 1 = 7.$$

Thus:

$$C_2 = 2 \quad \text{and} \quad C_1 = 2 .$$

Therefore via (4.44) the specific solution is:

$$x_n = C_1 + C_2 3^n - n = 2 + 2(3^n) - n \ , \quad n = 0, 1, \ldots .$$

We will advance our studies of second order nonhomogeneous and nonautonomous difference equation in the form:

$$x_{n+2} + px_{n+1} + qx_n = a_n \ , \quad n = 0, 1, \ldots ,$$

where $p \in \Re$, $q \neq 0$, $x_0, x_1 \in \Re$, and $\{a_n\}_{n=0}^{\infty}$ is a sequence of real numbers. The next two sections will consider two cases when:

(i) $\{a_n\}_{n=0}^{\infty} = r^n$, where $r \neq 0$ and $r \neq 1$.

(ii) $\{a_n\}_{n=0}^{\infty} = n^k$, for $k \in \mathbb{N}$.

4.4 Nonhomogeneous Second Order Linear Difference Equations with a Variable Geometric Coefficient

These are examples of second order linear nonhomogeneous and nonautonomous difference equations with a variable geometric coefficient:

(i) $x_{n+2} = 5x_{n+1} - 6x_n + 2^n$, $n = 0, 1, \ldots$.

(ii) $x_{n+2} = 4x_n + \left(\frac{-3}{4}\right)^n$, $n = 0, 1, \ldots$.

Now we will examine the second order nonhomogeneous and nonautonomous linear Δ.E. in the form:

$$x_{n+2} + px_{n+1} + qx_n = r^n \ , \quad n = 0, 1, \ldots , \tag{4.45}$$

where $x_0, x_1, p \in \Re$, $q, r \neq 0$, and $r \neq 1$. Similar to the previous section, via (4.3), (4.4), and (4.5), we let

$$x_n^h \ , \quad n = 0, 1, \ldots ,$$

be the **homogeneous solution** to Equation (4.45). Also, recall that Equation (4.2) resembles the characteristic polynomial of the homogeneous solution with two roots λ_1 and λ_2. Our next goal is to determine the **Particular Solution**

$$x_n^P \ , \quad n = 0, 1, \ldots ,$$

of Equation (4.45). The **Particular Solution** will depend on the relationship between r, λ_1, and λ_2 and will guide us to the following three cases:

CASE 1. Suppose that $r \neq \lambda_1$ and $r \neq \lambda_2$. Then:

$$x_n^P = A(r^n) \ , \quad n = 0, 1, \ldots . \tag{4.46}$$

To solve for A, we substitute Equation (4.46) into Equation (4.45).

CASE 2. Suppose that $r = \lambda_1$ and $r \neq \lambda_2$. Then:

$$x_n^P = An(r^n) \ , \quad n = 0, 1, \ldots . \tag{4.47}$$

To solve for A, we substitute Equation (4.47) into Equation (4.45).

CASE 3. Suppose that $r = \lambda_1 = \lambda_2$. Then:

$$x_n^P = An^2(r^n) \ , \quad n = 0, 1, \ldots . \tag{4.48}$$

To solve for A, we substitute Equation (4.48) into Equation (4.45).

The following examples will illustrate the implementation of the three cases above.

Example 10. Solve the Δ.E.:

$$x_{n+2} - 3x_{n+1} + 2x_n = 3^n \ , \quad n = 0, 1, \dots , \tag{4.49}$$

and verify that the solution is correct.

Solution: Via (4.3):

$$x_n^h = C_1 1^n + C_2 2^n = C_1 + C_2 2^n \ , \quad n = 0, 1, \dots . \tag{4.50}$$

To determine the Particular Solution, set

$$x_n^P = A(3^n) \ , \quad n = 0, 1, \dots . \tag{4.51}$$

By substituting (4.51) into Equation (4.49) gives us:

$$x_{n+2} - 3x_{n+1} + 2x_n$$

$$= 9A(3^n) - 3[3A(3^n)] + 2A3^n = 3^n \ ,$$

and

$$A = \frac{1}{2} .$$

Therefore:

$$x_n^P = A(3^n) = \frac{3^n}{2} \ , \quad n = 0, 1, \dots . \tag{4.52}$$

Thus the general solution of Equation (4.49) is:

$$x_n = x_n^h + x_n^P = C_1 + C_2 2^n + \frac{3^n}{2} \ , \quad n = 0, 1, \dots . \tag{4.53}$$

To check the general solution, it suffices to check the homogeneous and the particular solutions separately. First substitute (4.52) into Equation (4.49):

$$x_{n+2} - 3x_{n+1} + 2x_n$$

$$= \frac{9(3^n)}{2} - 3\left[\frac{3(3^n)}{2}\right] + 2\left[\frac{3^n}{2}\right] = 3^n .$$

Now substitute (4.50) into Equation (4.49):

$$x_{n+2} - 3x_{n+1} + 2x_n$$

$$= [C_1 + 4C_2 2^n] - 3[C_1 + 2C_2 2^n] + 2[C_1 + C_2 2^n]$$

$$= C_1 + 4C_2 2^n - 3C_1 - 6C_2 2^n + 2C_1 + 2C_2 2^n = 0.$$

The result follows.

Example 11. Solve the Δ.E.:

$$x_{n+2} - 5x_{n+1} + 6x_n = 2^n , \quad n = 0, 1, \dots , \tag{4.54}$$

and verify that the solution is correct.

Solution: Via (4.3):

$$x_n^h = C_1 2^n + C_2 3^n , \quad n = 0, 1, \dots . \tag{4.55}$$

To determine the Particular Solution, set:

$$x_n^P = An(2^n) , \quad n = 0, 1, \dots . \tag{4.56}$$

First substitute (4.56) into Equation (4.54) and we get:

$$x_{n+2} - 5x_{n+1} + 6x_n$$

$$= 4A(n+2)(2^n) - 5[2A(n+1)(2^n)] + 6An2^n = 2^n .$$

Hence

$$4An + 8A - 10An - 10A + 6An = 1 ,$$

and

$$A = -\frac{1}{2} .$$

Therefore:

$$x_n^P = An(2^n) = -\frac{n2^n}{2} , \quad n = 0, 1, \dots . \tag{4.57}$$

The general solution of Equation (4.54) is:

$$x_n = x_n^h + x_n^P = C_1 2^n + C_2 3^n - \frac{n2^n}{2} , \quad n = 0, 1, \dots . \tag{4.58}$$

To check the general solution, we will check the homogeneous and the particular solutions separately. First substitute (4.57) into Equation (4.54):

$$x_{n+2} - 5x_{n+1} + 6x_n$$

$$= \frac{-4(n+2)(2^n)}{2} - 5\left[\frac{-2(n+1)(2^n)}{2}\right] + 6\left[\frac{-n2^n}{2}\right] = 2^n$$

$$= -2(n+2)(2^n) + 5(n+1)(2^n) - 3n2^n = 2^n .$$

Now substitute (4.55) into Equation (4.54):

$$x_{n+2} - 5x_{n+1} + 6x_n = [4C_1 2^n + 9C_2 3^n] - 5[2C_1 2^n + 3C_2 3^n] + 6[C_1 2^n + C_2 3^n] = 0.$$

The result follows.

Example 12. Solve the Δ.E.:

$$x_{n+2} - 6x_{n+1} + 9x_n = 3^n \ , \quad n = 0, 1, \ldots , \tag{4.59}$$

and verify that the solution is correct.

Solution: Via (4.4):

$$x_n^h = C_1 3^n + C_2 n 3^n \ , \quad n = 0, 1, \ldots . \tag{4.60}$$

To determine the Particular Solution, set:

$$x_n^P = An^2(3^n) \ , \quad n = 0, 1, \ldots . \tag{4.61}$$

By substituting (4.61) into Equation (4.59) we obtain:

$$x_{n+2} - 6x_{n+1} + 9x_n = 9A(n+2)^2(3^n) - 6[3A(n+1)^2(3^n)] + 9An^2 3^n = 3^n .$$

Thus:

$$9A(n+2)^2 - 18A(n+1)^2 + 9An^2 = 9A\left[(n+2)^2 - 2(n+1)^2 + n^2\right] = 1 ,$$

and $A = \frac{1}{18}$. Therefore:

$$x_n^P = An^2(3^n) = \frac{n^2 3^n}{18} \ , \quad n = 0, 1, \ldots . \tag{4.62}$$

The general solution Equation (4.59) is:

$$x_n = x_n^h + x_n^P = C_1 3^n + C_2 n 3^n + \frac{n^2 3^n}{18} \ , \quad n = 0, 1, \ldots . \tag{4.63}$$

To check the general solution, we check the homogeneous and the particular solutions separately. First substitute (4.62) into Equation (4.59):

$$x_{n+2} - 6x_{n+1} + 9x_n = \frac{9(n+2)^2 3^n}{18} - 6\left[\frac{3(n+1)^2 3^n}{18}\right] + 9\left[\frac{n^2 3^n}{18}\right]$$

$$= \frac{3^n}{2}\left[(n+2)^2 - 2(n+1)^2 + n^2\right] = 3^n .$$

Now substitute (4.60) into Equation (4.59):

$$x_{n+2} - 6x_{n+1} + 9x_n$$

$$= [9C_1 3^n + 9C_2(n+2)3^n] - 6[3C_1 3^n + 3C_2(n+1)3^n] + 9[C_1 3^n + C_2 n 3^n]$$

$$= 0.$$

4.5 Nonhomogeneous Second Order Linear Difference with a Variable Coefficient n^k

These are examples of nonhomogeneous second order linear difference equations with a variable coefficient n^k:

(i) $x_{n+2} = 6x_{n+1} - 8x_n + n^3$, $n = 0, 1, \ldots$.

(ii) $x_{n+2} = x_n + n^2$, $n = 0, 1, \ldots$.

We will examine the second order linear nonhomogeneous Δ.E. in the form:

$$x_{n+2} + px_{n+1} + qx_n = n^k , \quad n = 0, 1, \ldots , \tag{4.64}$$

where $x_0, x_1, p \in \Re$, $q \neq 0$ and $k \in \mathbb{N}$. We will assume that $1 + p + q = 0$ and set:

$$x_n^P = A_{k+1}n^{k+1} + A_k n^k + \ldots + A_1 n , \quad n = 0, 1, \ldots . \tag{4.65}$$

To solve for the coefficients $A_{k+1}, A_k, \ldots, A_1$, we substitute Equation (4.65) into Equation (4.64). The following examples will illustrate solving Equation (4.65) using Equation (4.64).

Example 13. Solve the Δ.E.:

$$x_{n+2} - 4x_{n+1} + 3x_n = n , \quad n = 0, 1, \ldots , \tag{4.66}$$

and verify that the solution is correct.

Solution: Via (4.3), the homogeneous solution is

$$x_n^h = C_1 1^n + C_2 (3)^n = C_1 + C_2 (3)^n , \quad n = 0, 1, \ldots . \tag{4.67}$$

To determine the Particular Solution, set:

$$x_n^P = A_2 n^2 + A_1 n , \quad n = 0, 1, \ldots . \tag{4.68}$$

Now substitute (4.68) into Equation (4.66):

$$x_{n+2} - 4x_{n+1} + 3x_n$$

$$= \left[A_2(n+2)^2 + A_1(n+2) \right] - 4\left[A_2(n+1)^2 + A_1(n+1) \right] + 3\left[A_2 n^2 + A_1 n \right] = n .$$

To solve for A_1 and A_2, we set $n = 0$ and $n = 1$ into above equation and we obtain:

$$A_1 = 0 \ (when \ n = 0) ,$$

$$-4A_2 - 2A_1 = 1 \ (when \ n = 1) ,$$

with two solutions

$$A_2 = -\frac{1}{4} \quad \text{and} \quad A_1 = 0.$$

Therefore:

$$x_n^P = A_2 n^2 + A_1 n = -\frac{n^2}{4}, \quad n = 0, 1, \dots. \tag{4.69}$$

The general solution of Equation (4.66) is:

$$x_n = x_n^h + x_n^P = C_1 + C_2(3)^n - \frac{n^2}{4}, \quad n = 0, 1, \dots. \tag{4.70}$$

To check the general solution, we check the homogeneous and the particular solutions separately. First by substituting (4.69) into Equation (4.66):

$$x_{n+2} - 4x_{n+1} + 3x_n$$

$$= \left[-\frac{(n+2)^2}{4} \right] - 3 \left[-\frac{(n+1)^2}{4} \right] + 4 \left[-\frac{n^2}{4} \right]$$

$$= \frac{1}{4} \left[-(n+2)^2 + 3(n+1)^2 - 4n^2 \right] = n.$$

Now substitute (4.67) into Equation (4.66):

$$x_{n+2} - 4x_{n+1} + 3x_n$$

$$= [C_1 + 9C_2(3)^n] - 4[C_1 + 3C_2(3)^n] + 3[C_1 + 3C_2(3)^n] = 0.$$

Now we will proceed with analyzing periodic traits of nonautonomous second order linear difference equations.

4.6 Periodic Solutions of Second Order Homogeneous Linear Difference Equations

Our plan is to show that Equation (4.1) exhibits periodic solutions with any period-p, $(p \geq 2)$. First of all, recall that every nontrivial solution of:

$$x_{n+2} = x_n, \quad n = 0, 1, \dots,$$

is periodic with period-2. Second, recall that Equation (4.1) has periodic solutions with period-p, $(p \geq 3)$ when the characteristic polynomial of Equation (4.1) has two imaginary complex conjugate roots:

$$\lambda_1 = A + Bi, \quad \lambda_2 = A - Bi,$$

and

$$\Lambda = |A + Bi| = |A - Bi| = 1.$$

In this case, solution of Equation (4.1) is:

$$x_n = C_1 sin(n\theta) + C_2 cos(n\theta) \quad , \quad n = 0, 1, \dots .$$

(4.71)

Equation (4.71) is periodic with period-p, $p \geq 3$, where $\theta = \frac{2\pi}{p}$, and either $\theta = arg(\lambda_1)$ or $\theta = arg(\lambda_2)$. Furthermore, Equation (4.71) is a solution of:

$$x_{n+2} = 2cos\left(\frac{2\pi}{p}\right) x_{n+1} - x_n ,$$

(4.72)

where every nontrivial solution is periodic with period-p, $(p \geq 3)$. The following examples describe various periodic patterns.

Example 14. Determine the period-2 pattern of:

$$x_{n+2} = x_n \quad , \quad n = 0, 1, \dots .$$

(4.73)

Solution: By iteration we obtain:

$$x_0 ,$$

$$x_1 ,$$

$$x_2 = x_0 ,$$

$$x_3 = x_1 ,$$

and the following period-2 pattern:

$$x_0 , x_1 , x_0 , x_1, \dots .$$

Hence every nontrivial solution is a period-2 solution.

Example 15. Determine the period-3 pattern of:

$$x_{n+2} = -x_{n+1} - x_n \quad , \quad n = 0, 1, \dots .$$

(4.74)

Solution: We acquire:

$$x_0 ,$$

$$x_1 ,$$

$$x_2 = -x_1 - x_0 ,$$

$$x_3 = - [x_2] - x_1 = x_0 + x_1 - x_1 = x_0 ,$$

$$x_4 = - [x_3] - [x_2] = -x_0 + x_1 + x_0 = x_1 ,$$

and the following period-3 pattern:

$$x_0 \, , \, x_1 \, , \, -x_1 - x_0 \, , \, x_0 \, , \, x_1 \, , \, -x_1 - x_0 \, , \, \dots \, .$$

Observe that every nontrivial solution is a period-3 solution. The characteristic polynomial of Equation (4.74) is:

$$\lambda^2 + \lambda + 1 = 0 \, ,$$

whose roots are $\lambda_1 = -\frac{1}{2} + \frac{\sqrt{3}}{2}i$ and $\lambda_2 = -\frac{1}{2} - \frac{\sqrt{3}}{2}i$. Notice that $\theta = arg(\lambda_1) = \frac{2\pi}{3}$ and the result follows.

Example 16. Determine the period-4 pattern of:

$$x_{n+2} = -x_n \, , \quad n = 0, 1, \dots \, . \tag{4.75}$$

Solution: By iteration we procure the following period-4 pattern:

$$x_0 \, , \, x_1 \, , \, -x_0 \, , \, -x_1 \, , \, \dots \, .$$

Every nontrivial solution portrays an alternating period-4 pattern. The characteristic polynomial of Equation (4.75) is:

$$\lambda^2 + 1 = 0 \, ,$$

whose roots are $\lambda_1 = i$ and $\lambda_2 = -i$ and $\theta = arg(\lambda_1) = \frac{\pi}{2}$. The result follows.

Example 17. Determine the period-6 pattern of:

$$x_{n+2} = x_{n+1} - x_n \, , \quad n = 0, 1, \dots \, . \tag{4.76}$$

Solution: We obtain the following period-6 pattern:

$$x_0 \, , \, x_1 \, , \, x_1 - x_0 \, , \, -x_0 \, , \, -x_1 \, , \, x_0 - x_1 \, , \, \dots \, .$$

Every nontrivial evokes an alternating period-6 pattern. The characteristic polynomial of Equation (4.76) is:

$$\lambda^2 - \lambda + 1 = 0 \, ,$$

whose roots are $\lambda_1 = \frac{1}{2} + \frac{\sqrt{3}}{2}i$ and $\lambda_2 = \frac{1}{2} - \frac{\sqrt{3}}{2}i$ and $\theta = arg(\lambda_1) = \frac{\pi}{3}$. The result follows.

From Examples (15)–(17), even ordered periodic cycles exhibit alternating patterns. By induction we can prove that every nontrivial solution of Equation (4.72) is periodic with period-p, $(p \geq 3)$. The characteristic polynomial of Equation (4.72) is:

$$\lambda^2 - 2cos\left(\frac{2\pi}{p}\right)\lambda + 1 = 0 \, ,$$

with two imaginary roots λ_1 and λ_2 and either $\theta = \frac{2\pi}{p} = arg(\lambda_1)$ or $\theta = \frac{2\pi}{p} = arg(\lambda_2)$.

4.7 Periodic Solutions of Nonautonomous Second Order Linear Difference Equations

Suppose $\{a_n\}_{n=0}^{\infty}$ is a period-k sequence, $(k \geq 2)$. First of all, our aim is to investigate the existence and patterns of the periodic cycles of nonautonomous second order linear Δ.E. in the form:

$$x_{n+2} = a_n x_n \quad , \quad n = 0, 1, \dots , \tag{4.77}$$

where $x_0 \neq 0$. The following theorem outlines the periodic traits of Equation (4.77).

Theorem 1. *Suppose that $\{a_n\}_{n=0}^{\infty}$ is a period-k sequence, $(k \geq 2)$ and $x_0 \neq 0$. Then the following statements are true:*

(i) *If $\{a_n\}_{n=0}^{\infty}$ is a period-2 sequence, then Equation (4.77) has no period-2 cycles.*

(ii) *If $\{a_n\}_{n=0}^{\infty}$ is a period-2 sequence, then every solution of Equation (4.77) is periodic with period-4 if and only if either $a_0 = \pm 1$ and $a_1 = \pm 1$ and $a_0 \neq a_1$.*

(iii) *If $\{a_n\}_{n=0}^{\infty}$ is a period-2k sequence $(k \geq 2)$, then every solution of Equation (4.77) is periodic with period-2k if and only if:*

$$\prod_{i=0}^{k-1} a_{2i} = 1 \quad \text{and} \quad \prod_{i=0}^{k-1} a_{2i+1} = 1 .$$

(iv) *If $\{a_n\}_{n=0}^{\infty}$ is a period-$(2k+1)$ sequence $(k \in \mathbb{N})$, then every solution of Equation (4.77) is periodic with period-$2(2k+1)$ if and only if:*

$$\prod_{i=0}^{2k} a_i = 1 .$$

Theorem (1) is proved by induction and will be left as an exercise at the end of the chapter. Our next objective is to determine the necessary and sufficient conditions for the existence and uniqueness of periodic cycles of:

$$x_{n+2} = \left[2\cos\left(\frac{2\pi}{p}\right) x_{n+1} - x_n \right] + a_n \quad , \quad n = 0, 1, \dots . \tag{4.78}$$

Recall that every nontrivial solution of Equation (4.72) is periodic with period-p. The first three examples will focus on periodicity of:

$$x_{n+2} = x_n + a_n \quad , \quad n = 0, 1, \dots . \tag{4.79}$$

Recall that every nontrivial solution of Equation (4.73) is periodic with period-2. The following three examples will assume that $\{a_n\}_{n=0}^{\infty}$ is a period-2k sequence $(k \in \mathbb{N})$ and will examine the periodic traits of Equation (4.79).

Example 18. Suppose that $\{a_n\}_{n=0}^{\infty}$ is a period-2 sequence. Show that:

$$x_{n+2} = x_n + a_n \ , \quad n = 0, 1, \ldots \ ,$$

has no period-2 cycles and explain why.

Solution: Notice:

$$x_0 \ ,$$

$$x_1 \ ,$$

$$x_2 = x_0 + a_0 \ ,$$

$$x_3 = x_1 + a_1 \ ,$$

$$x_4 = [x_2] + a_0 = x_0 + a_0 + a_0 = x_0 + 2a_0 \ ,$$

$$x_5 = [x_3] + a_1 = x_1 + a_1 + a_1 = x_1 + 2a_1 \ .$$

The existence of a period-2 cycle is possible only when $a_0 = 0$ and $a_1 = 0$, which is clearly a contradiction as we assumed that $\{a_n\}_{n=0}^{\infty}$ is a period-2 sequence. Every solution of Equation (4.73) is periodic with period-2. However, it is impossible to obtain a period-2 cycle as a_0 and a_1 do not cancel. Furthermore, we conclude that it is impossible to produce any periodic cycles when $\{a_n\}_{n=0}^{\infty}$ is a period-2 sequence.

Example 19. Suppose that $\{a_n\}_{n=0}^{\infty}$ is a period-4 sequence. Determine the necessary and sufficient conditions for the existence of period-4 cycles of:

$$x_{n+2} = x_n + a_n \ , \quad n = 0, 1, \ldots \ .$$

Solution: Observe:

$$x_0 \ ,$$

$$x_1 \ ,$$

$$x_2 = x_0 + a_0 \ ,$$

$$x_3 = x_1 + a_1 \ ,$$

$$x_4 = [x_2] + a_2 = x_0 + a_0 + a_2 = x_0 \ ,$$

$$x_5 = [x_3] + a_3 = x_1 + a_1 + a_3 = x_1 \ .$$

Every solution is periodic with period-4 if and only if:

$$a_0 + a_2 = 0 \quad \text{and} \quad a_1 + a_3 = 0 \ .$$

Every solution of Equation (4.73) is periodic with period-2. Since $\{a_n\}_{n=0}^{\infty}$ is a period-4 sequence, we cannot acquire period-2 cycles.

Example 20. Suppose that $\{a_n\}_{n=0}^{\infty}$ is a period-6 sequence. Determine the necessary and sufficient conditions for the existence of period-6 solutions of:

$$x_{n+2} = x_n + a_n \ , \quad n = 0, 1, \ldots .$$

Solution: Similar to the previous example, we show that every solution is periodic with period-6 if and only if:

$$a_0 + a_2 + a_4 = 0 \ \text{ and } \ a_1 + a_3 + a_5 = 0 .$$

Since every solution of Equation (4.73) is periodic with period-2 and as $\{a_n\}_{n=0}^{\infty}$ is a period-6 sequence, then we cannot exhibit period-2 cycles but exhibit period-6 cycles instead.

From Examples (24)-(29) we conclude that Equation (4.79) has period-2k cycles, $(k \geq 2)$ and the following theorem outlines the result.

Theorem 2. *Suppose that $\{a_n\}_{n=0}^{\infty}$ is a period-2k sequence, $(k \geq 2)$. Then every solution of:*

$$x_{n+2} = x_n + a_n \ , \quad n = 0, 1, \ldots ,$$

is periodic with period-2k, $(k \geq 2)$ if and only if:

$$\sum_{i=0}^{k-1} a_{2i} = 0 \ \text{ and } \ \sum_{i=0}^{k-1} a_{2i+1} = 0 .$$

Proof: *Similar to Examples (24) and (29), we obtain:*

$$x_0 ,$$

$$x_1 ,$$

$$x_2 = x_0 + a_0 ,$$

$$x_3 = x_1 + a_1 ,$$

$$x_4 = x_2 + a_2 = x_0 + [a_0 + a_2] ,$$

$$x_5 = x_3 + a_3 = x_1 + [a_1 + a_3] ,$$

$$x_6 = x_4 + a_4 = x_0 + [a_0 + a_2 + a_4] ,$$

$$x_7 = x_5 + a_5 = x_1 + [a_1 + a_3 + a_5] ,$$

$$x_8 = x_6 + a_6 = x_0 + [a_0 + a_2 + a_4 + a_6] \, ,$$

$$x_9 = x_7 + a_7 = x_1 + [a_1 + a_3 + a_5 + a_7] \, .$$

$$\vdots$$

Then:

$$\begin{cases} x_{2k} = x_0 + \sum_{i=0}^{k-1} a_{2i} \, , \\[2em] x_{2k+1} = x_1 + \sum_{i=0}^{k-1} a_{2i+1} \, . \end{cases}$$

The result follows.

The case when $\{a_n\}_{n=0}^{\infty}$ is periodic with an odd period-$(2k+1)$, $(k \in \mathbb{N})$ will be left as an exercise at the end of the chapter to investigate by addressing the following two questions: Do periodic cycles exist? What is the period (even order or odd order)? Now we will shift our focus on the periodic traits of:

$$x_{n+2} = -[x_{n+1} - x_n] + a_n \, , \quad n = 0, 1, \dots . \tag{4.80}$$

Recall that every nontrivial solution of Equation (4.74) is periodic with period-3. The next two examples will assume that $\{a_n\}_{n=0}^{\infty}$ is a period-3k sequence ($k \in \mathbb{N}$) and will examine the periodic properties of Equation (4.80).

Example 21. Suppose that $\{a_n\}_{n=0}^{\infty}$ is a period-3 sequence. Show that:

$$x_{n+2} = -[x_{n+1} + x_n] + a_n \, , \quad n = 0, 1, \dots ,$$

has no period-3 solutions and explain why.

Solution: Observe:

$$x_0 \, ,$$

$$x_1 \, ,$$

$$x_2 = -x_1 - x_0 + a_0 \, ,$$

$$x_3 = -[x_2] - x_1 + a_1 = x_0 - a_0 + a_1 = x_0 \, ,$$

$$x_4 = -[x_3] - [x_2] + a_2 = x_1 - a_1 + a_2 = x_1 \, .$$

A period-3 cycle is possible only when

$$a_0 = a_1 = a_2 \, ,$$

which is clearly a contradiction as we assumed that $\{a_n\}_{n=0}^{\infty}$ is a period-3 sequence. On one hand, every solution of Equation (4.74) is period with period-3. On the other hand, it is impossible to obtain a period-3 cycle as it contradicts that $\{a_n\}_{n=0}^{\infty}$ is a period-3 sequence. This similar phenomenon occurred in Example (18).

Example 22. Suppose that $\{a_n\}_{n=0}^{\infty}$ is a period-6 sequence. Determine the necessary and sufficient conditions for the existence of period-6 solutions of:

$$x_{n+2} = -[x_{n+1} + x_n] + a_n \quad , \quad n = 0, 1, \dots .$$

Solution: Notice:

$$x_0 ,$$

$$x_1 ,$$

$$x_2 = -x_1 - x_0 + a_0 ,$$

$$x_3 = -[x_2] - x_1 + a_1 = x_0 - a_0 + a_1 ,$$

$$x_4 = -[x_3] - [x_2] + a_2 = x_1 - a_1 + a_2 ,$$

$$x_5 = -[x_4] - [x_3] + a_3 = -x_0 - x_1 + a_0 - a_2 + a_3 ,$$

$$x_6 = -[x_5] - [x_4] + a_4 = x_0 - a_0 - a_3 + a_1 + a_4 = x_0 ,$$

$$x_7 = -[x_6] - [x_5] + a_5 = x_1 - a_0 - a_2 - a_3 + a_5 = x_1 .$$

Period-6 cycle is possible only when:

$$a_0 + a_3 = a_1 + a_4 = a_2 + a_5 .$$

Every nontrivial solution of Equation (4.74) is periodic with period-3. As $\{a_n\}_{n=0}^{\infty}$ is a period-6 sequence, we cannot acquire a period-3 cycle.

From Example 22, following theorem outlines the result when $\{a_n\}_{n=0}^{\infty}$ is a period-3k sequence, $(k \geq 2)$.

Theorem 3. *Suppose that* $\{a_n\}_{n=0}^{\infty}$ *is a period-3k sequence, $(k \geq 2)$. Then every solution of:*

$$x_{n+2} = -[x_{n+1} + x_n] + a_n \quad , \quad n = 0, 1, \dots ,$$

is periodic with period-3k, $(k \geq 2)$ if and only if:

$$\sum_{i=1}^{k} a_{3i-3} = \sum_{i=1}^{k} a_{3i-2} = \sum_{i=1}^{k} a_{3i-1} .$$

Theorem (3) is proved by induction and will be left as an exercise at the end of the chapter. The next pertinent question to address: will periodic solutions exist if

$\{a_n\}_{n=0}^{\infty}$ is a period-p sequence where p is not a multiple of 3? Several computer observations will be required to answer this question. The next sequential examples will focus on the periodic nature of solutions of:

$$x_{n+2} = -[x_n] + a_n \ , \quad n = 0, 1, \dots , \tag{4.81}$$

Recall that every nontrivial solution of Equation (4.75) is periodic with period-4.

Example 23. Suppose that $\{a_n\}_{n=0}^{\infty}$ is a period-2 sequence. Show that every solution of:

$$x_{n+2} = -[x_n] + a_n \ , \quad n = 0, 1, \dots ,$$

is periodic with period-4.

Solution: Observe that:

$$x_0 \ ,$$

$$x_1 \ ,$$

$$x_2 = -x_0 + a_0 \ ,$$

$$x_3 = -x_1 + a_1 \ ,$$

$$x_4 = -[x_2] + a_0 = x_0 - a_0 + a_0 = x_0 \ ,$$

$$x_5 = -[x_3] + a_1 = x_1 - a_1 + a_1 = x_1 \ ,$$

describe the following period-4 pattern:

$$x_0 \ , \ x_1 \ , \ -x_0 + a_0 \ , \ -x_1 + a_1, \ \dots \ .$$

Every nontrivial solution of Equation (4.75) is periodic with period-4. In addition, we see that every solution of Equation (4.81) is also periodic with period-4. On the other hand, we see that the period of the sequence $\{a_n\}_{n=0}^{\infty}$ is shorter than the period of Equation (4.75). The immediate question to address: is it possible to obtain a period-2 solution? The next example will address the answer to this question.

Example 24. Suppose that $\{a_n\}_{n=0}^{\infty}$ is a period-2 sequence. Determine the unique period-2 solution of:

$$x_{n+2} = -[x_n] + a_n \ , \quad n = 0, 1, \dots \ .$$

Solution: Set $x_2 = x_0$ and $x_3 = x_1$ and:

$$x_0 \, ,$$

$$x_1 \, ,$$

$$x_2 = -x_0 + a_0 = x_0 \, ,$$

$$x_3 = -x_1 + a_1 = x_1 \, .$$

Solving for x_0 and x_1 gives us the unique period-2 pattern:

$$\frac{a_0}{2} \, , \, \frac{a_1}{2} \, , \, \ldots \, .$$

Example 25. Suppose that $\{a_n\}_{n=0}^{\infty}$ is a period-4 sequence. Show that:

$$x_{n+2} = -[x_n] + a_n \quad , \quad n = 0, 1, \ldots \, ,$$

has no period-4 solutions and explain why.

Solution: Notice:

$$x_0 \, ,$$

$$x_1 \, ,$$

$$x_2 = -x_0 + a_0 \, ,$$

$$x_3 = -x_1 + a_1 \, ,$$

$$x_4 = -x_2 + a_2 = x_0 - a_0 + a_2 = x_0 \, ,$$

$$x_5 = -x_3 + a_3 = x_0 - a_1 + a_3 = x_1 \, .$$

Period-4 cycles are possible only when $a_0 = a_2$ and $a_1 = a_3$, which is clearly a contradiction as we assumed that $\{a_n\}_{n=0}^{\infty}$ is a period-4 sequence. Every solution of Equation (4.75) is period with period-4. On the other hand, period-4 cycles are impossible as $a_0 = a_2$ and $a_1 = a_3$ contradict that $\{a_n\}_{n=0}^{\infty}$ is a period-4 sequence. This phenomenon happened in Examples (18) and (21).

Example 26. Suppose that $\{a_n\}_{n=0}^{\infty}$ is a period-8 sequence. Determine the necessary and sufficient conditions for the existence of period-8 solutions of:

$$x_{n+2} = -[x_n] + a_n \quad , \quad n = 0, 1, \ldots \, .$$

Solution: Similar to the previous examples, every solution is periodic with period-8 if and only if:

$$a_0 + a_4 = a_2 + a_6 \quad \text{and} \quad a_1 + a_5 = a_1 + a_7 \, .$$

Every solution Equation (4.75) is periodic with period-4 and as $\{a_n\}_{n=0}^{\infty}$ is a period-8 sequence, then period-4 cycles are not possible.

From Example 26, following theorem describes the result when $\{a_n\}_{n=0}^{\infty}$ is a period-4k sequence, $(k \geq 2)$.

Theorem 4. *Suppose that $\{a_n\}_{n=0}^{\infty}$ is a period-4k sequence, $(k \geq 2)$. Then every solution of:*

$$x_{n+2} = -[x_n] + a_n \quad , \quad n = 0, 1, \ldots ,$$

is periodic with period-4k, $(k \geq 2)$ if and only if:

$$\sum_{i=1}^{k} a_{4i-4} = \sum_{i=1}^{k} a_{4i-2} \text{ and } \sum_{i=1}^{k} a_{4i-3} = \sum_{i=1}^{k} a_{4i-1}.$$

The proof of Theorem (4) will be left as an exercise. Can periodic solutions exist when $\{a_n\}_{n=0}^{\infty}$ is a period-k sequence, where k is not a multiple of 4? The next four examples will focus on the periodic nature of the following Δ.E.:

$$x_{n+2} = [x_{n+1} - x_n] + a_n \quad , \quad n = 0, 1, \ldots . \tag{4.82}$$

Recall that every nontrivial solution of Equation (4.76) is periodic with period-6.

Example 27. Suppose that $\{a_n\}_{n=0}^{\infty}$ is a period-3 sequence. Show that every solution of:

$$x_{n+2} = [x_{n+1} - x_n] + a_n \quad , \quad n = 0, 1, \ldots ,$$

is periodic with period-6.

Solution: Similar to Example (23), we acquire:

$$x_0 ,$$

$$x_1 ,$$

$$x_2 = x_1 - x_0 + a_0 ,$$

$$x_3 = x_2 - x_1 + a_1 = -x_0 + a_0 + a_1 ,$$

$$x_4 = x_3 - x_2 + a_2 = -x_1 + a_1 + a_2 ,$$

$$x_5 = x_4 - x_3 + a_0 = x_0 - x_1 + a_2 ,$$

$$x_6 = x_5 - x_4 + a_1 = x_0 ,$$

$$x_7 = x_6 - x_5 + a_2 = x_1 .$$

The result follows.

Can Equation (4.82) have period-3 cycles when $\{a_n\}_{n=0}^{\infty}$ is a period-3 sequence? The next example will remit the answer to this question.

Example 28. Suppose that $\{a_n\}_{n=0}^{\infty}$ is a period-3 sequence. Determine the unique period-3 solution of:

$$x_{n+2} = -[x_n] + a_n \quad , \quad n = 0, 1, \dots .$$

Solution: Setting $x_3 = x_0$ and $x_4 = x_1$ gives us:

$$x_0 ,$$

$$x_1 ,$$

$$x_2 = x_1 - x_0 + a_0 ,$$

$$x_3 = x_2 - x_1 + a_1 = -x_0 + a_0 + a_1 = x_0 ,$$

$$x_4 = x_3 - x_2 + a_2 = -x_1 + a_1 + a_2 = x_1 .$$

By solving for x_0 and x_1, we procure the following unique period-3 pattern:

$$\frac{a_0 + a_1}{2} , \frac{a_1 + a_2}{2} , \frac{a_2 + a_0}{2} \dots .$$

Example 29. Suppose that $\{a_n\}_{n=0}^{\infty}$ is a period-6 sequence. Show that:

$$x_{n+2} = [x_{n+1} - x_n] + a_n \quad , \quad n = 0, 1, \dots ,$$

has no period-6 solutions and explain why.

Solution: Similar to Examples (18), (21) and (23), period-6 cycles are possible only when $a_0 = a_3$, $a_1 = a_4$, and $a_2 = a_5$, which is clearly a contradiction as we assumed that $\{a_n\}_{n=0}^{\infty}$ is a period-6 sequence.

Example 30. Suppose that $\{a_n\}_{n=0}^{\infty}$ is a period-12 sequence. Determine the necessary and sufficient conditions for the existence of period-12 solutions of:

$$x_{n+2} = [x_{n+1} - x_n] + a_n \quad , \quad n = 0, 1, \dots .$$

Solution: Similar to previous examples, we show that every solution is periodic with period-12 if and only if:

$$a_0 + a_6 = a_2 + a_8 = a_4 + a_{10} \quad \text{and} \quad a_1 + a_7 = a_3 + a_9 = a_5 + a_{11} .$$

From Example 30, following theorem extends the result when $\{a_n\}_{n=0}^{\infty}$ is a period-6k sequence, $(k \geq 2)$.

Theorem 5. *Suppose that $\{a_n\}_{n=0}^{\infty}$ is a period-6k sequence, $(k \geq 2)$. Then every solution of:*

$$x_{n+2} = [x_{n+1} - x_n] + a_n \quad, \quad n = 0, 1, \ldots,$$

is periodic with period-6k, $(k \geq 2)$ if and only if:

$$\sum_{i=1}^{k} a_{6i-6} = \sum_{i=1}^{k} a_{6i-4} = \sum_{i=1}^{k} a_{6i-2} \quad \text{and}$$

$$\sum_{i=1}^{k} a_{6i-5} = \sum_{i=1}^{k} a_{6i-3} = \sum_{i=1}^{k} a_{6i-1}.$$

The proof of Theorem (5) will be left as an exercise to prove at the end of the chapter. From Examples 21, 25 and 29, the following theorem generalizes the result when $\{a_n\}_{n=0}^{\infty}$ is a period-k sequence, $(k \geq 3)$.

Theorem 6. *Suppose that $\{a_n\}_{n=0}^{\infty}$ is a period-k sequence, $(k \geq 3)$. Then:*

$$x_{n+2} = \left[2\cos\left(\frac{2\pi}{k}\right)x_{n+1} - x_n\right] + a_n \quad, \quad n = 0, 1, \ldots,$$

has no periodic solutions with period-k.

Proving Theorem 6 will require two cases; when k is even and when k is odd. Furthermore, from Example 23 and Example 29, we will pose the following conjecture that can be verified by computer observations and analysis.

Conjecture 1. Suppose that $\{a_n\}_{n=0}^{\infty}$ is a period-k sequence, $(k \geq 3)$. Then every solution of:

$$x_{n+2} = \left[2\cos\left(\frac{2\pi}{k}\right)x_{n+1} - x_n\right] + a_n \quad, \quad n = 0, 1, \ldots,$$

is periodic with period-p if $p < k$, and $k = Np$ for some $N \geq 2$.

Furthermore, from Examples 24 and 28, we will pose the following conjecture:

Conjecture 2. Suppose that $\{a_n\}_{n=0}^{\infty}$ is a period-k sequence, $(k \geq 4)$. Then:

$$x_{n+2} = \left[2\cos\left(\frac{2\pi}{p}\right)x_{n+1} - x_n\right] + a_n \quad, \quad n = 0, 1, \ldots,$$

either:

(1) Has a unique periodic cycle with period-k, or

(2) Every solution is periodic with period-p,

where $k < p$, and $p = Nk$ for some $N \geq 2$.

Now we will suggest the following **Open Problem** that addresses the existence of specific periodic cycles.

Open Problem 7. *Suppose that $\{a_n\}_{n=0}^{\infty}$ is a period-k sequence, $(k \geq 2)$. Does theΔ.E.:*

$$x_{n+2} = \left[2\cos\left(\frac{2\pi}{p}\right)x_{n+1} - x_n\right] + a_n \quad, \quad n = 0, 1, \ldots :$$

exhibit periodic solutions when p is not a multiple of k? Very thorough computer observations will be required to answer this question. For example, will periodic solutions exist when $p = 3$ and $k = 2$?

Now we will shift our investigation of periodic cycles of the nonautonomous linear Δ.E.:

$$x_{n+2} = a_n x_n + b_n \quad, \quad n = 0, 1, \ldots, \tag{4.83}$$

where $\{a_n\}_{n=0}^{\infty}$ and $\{b_n\}_{n=0}^{\infty}$ are periodic sequences with either the same period or with different periods. Only unique periodic cycles of Equation (4.83) will appear.

Example 31. Suppose that $\{a_n\}_{n=0}^{\infty}$ and $\{b_n\}_{n=0}^{\infty}$ are period-2 sequences. Determine the unique period-2 pattern of:

$$x_{n+2} = a_n x_n + b_n \quad, \quad n = 0, 1, \ldots .$$

Solution: Setting $x_2 = x_0$ and $x_3 = x_1$ gives us:

$$x_0 \,,$$

$$x_1 \,,$$

$$x_2 = a_0 x_0 + b_0 = x_0 \,,$$

$$x_3 = a_1 x_1 + b_1 = x_1 \,.$$

By solving for x_0 and x_1 we procure the following unique period-2 pattern:

$$\frac{b_1}{1 - a_0} \,, \quad \frac{b_1}{1 - a_1} \,, \quad \frac{b_1}{1 - a_0} \,, \quad \frac{b_1}{1 - a_1} \,, \quad \ldots \,,$$

provided that $a_0 \neq 1$ and $a_1 \neq 1$. Observe that from neighbor to neighbor all the indices of both sequences $\{a_n\}_{n=0}^{\infty}$ and $\{b_n\}_{n=0}^{\infty}$ shift by one.

Example 32. Suppose that $\{a_n\}_{n=0}^{\infty}$ and $\{b_n\}_{n=0}^{\infty}$ are period-4 sequences. Determine the unique period-4 pattern of:

$$x_{n+2} = a_n x_n + b_n \quad, \quad n = 0, 1, \ldots .$$

Solution: Set $x_4 = x_0$ and $x_5 = x_1$ and we acquire:

$$x_0 ,$$

$$x_1 ,$$

$$x_2 = a_0 x_0 + b_0 ,$$

$$x_3 = a_1 x_1 + b_1 ,$$

$$x_4 = a_2 [x_2] + b_2 = a_0 a_2 x_0 + a_2 b_0 + b_2 = x_0 ,$$

$$x_5 = a_3 [x_3] + b_3 = a_1 a_3 x_0 + a_3 b_1 + b_3 = x_1 .$$

Solving for x_0 and x_1 gives us the following unique period-4 pattern:

$$\frac{a_2 b_0 + b_2}{1 - a_0 a_2} , \frac{a_3 b_1 + b_3}{1 - a_1 a_3} , \frac{a_0 b_2 + b_0}{1 - a_0 a_2} , \frac{a_1 b_3 + b_1}{1 - a_1 a_3} , \dots ,$$

provided that $a_0 a_2 \neq 1$ and $a_0 a_3 \neq 1$.

From Examples 31 and 32, the following theorem generalizes the result when $\{a_n\}_{n=0}^\infty$ and $\{b_n\}_{n=0}^\infty$ are periodic-2k sequences, $(k \geq 2)$.

Theorem 8. *Suppose that $\{a_n\}_{n=0}^\infty$ and $\{b_n\}_{n=0}^\infty$ are periodic sequences with period-2k, $(k \geq 2)$. Then:*

$$x_{n+2} = a_n x_n + b_n , \quad n = 0, 1, \dots ,$$

has a unique period-2k cycle if

$$\prod_{i=1}^k a_{2i-2} \neq 1 \quad \text{and} \quad \prod_{i=1}^k a_{2i-1} \neq 1 ,$$

and

$$x_0 = \frac{a_{2k-2} [\dots [a_4 [a_2 b_0 + b_2] + b_4] \dots] + b_{2k-2}}{1 - \left[\prod_{i=1}^k a_{2i-2}\right]} ,$$

$$x_1 = \frac{a_{2k-1} [\dots [a_5 [a_3 b_1 + b_3] + b_5] \dots] + b_{2k-1}}{1 - \left[\prod_{i=1}^k a_{2i-1}\right]} .$$

The proof of Theorem 8 will be left as an exercise at the end of the chapter. Furthermore, from Examples 31 and 32, we will present the following two **Open Problems**:

Open Problem 9. *Suppose that $\{a_n\}_{n=0}^\infty$ and $\{b_n\}_{n=0}^\infty$ are periodic sequences with period-$(2k+1)$, $(k \in \mathbb{N})$. Discuss the periodic character of Equation (4.83).*

Open Problem 10. *Suppose that $\{a_n\}_{n=0}^\infty$ is periodic with period-k, $(k \geq 2)$, and $\{b_n\}_{n=0}^\infty$ is periodic with period-l, $(l \geq 2)$ and $k \neq l$. Discuss the periodic character of Equation (4.83).*

Now we advance with investigating the periodic nature of solutions of the nonautonomous linear Δ.E.:

$$x_{n+2} = -a_n x_n + b_n \quad , \quad n = 0, 1, \dots , \tag{4.84}$$

where $\{a_n\}_{n=0}^{\infty}$ and $\{b_n\}_{n=0}^{\infty}$ are periodic sequences with either the same period or different periods. Only unique periodic cycles of Equation (4.84) will appear.

Example 33. Suppose that $\{a_n\}_{n=0}^{\infty}$ and $\{b_n\}_{n=0}^{\infty}$ are period-2 sequences. Determine the pattern of the unique period-2 cycle of:

$$x_{n+2} = -a_n x_n + b_n \quad , \quad n = 0, 1, \dots .$$

Solution: Setting $x_2 = x_0$ and $x_3 = x_1$ gives us:

$$x_0 ,$$

$$x_1 ,$$

$$x_2 = -a_0 x_0 + b_0 = x_0 ,$$

$$x_3 = -a_1 x_1 + b_1 = x_1 ,$$

Now we solve for x_0 and x_1 and we acquire the following period-2 pattern:

$$\frac{b_0}{a_0 + 1} , \frac{b_1}{a_1 + 1} , \frac{b_0}{a_0 + 1} , \frac{b_1}{a_1 + 1} , \dots .$$

First of all, we obtain a unique period-2 cycle. Second, notice that from neighbor to neighbor all the indices shift by one in both sequences $\{a_n\}_{n=0}^{\infty}$ and $\{b_n\}_{n=0}^{\infty}$.

Example 34. Suppose that $\{a_n\}_{n=0}^{\infty}$ and $\{b_n\}_{n=0}^{\infty}$ are period-4 sequences. Determine the pattern of the unique period-4 cycle of:

$$x_{n+2} = -a_n x_n + b_n \quad , \quad n = 0, 1, \dots .$$

Solution: We set $x_4 = x_0$ and $x_5 = x_1$ and we acquire:

$$x_0 ,$$

$$x_1 ,$$

$$x_2 = -a_0 x_0 + b_0 ,$$

$$x_3 = -a_1 x_1 + b_1 ,$$

$$x_4 = -a_2[x_2] + b_2 = -a_2[-a_0x_0 + b_0] + b_2 = x_0,$$

$$x_5 = -a_3[x_3] + b_3 = -a_3[-a_1x_1 + b_1] + b_3 = x_1.$$

Now we solve for x_0 and x_1 and we procure the following period-4 pattern:

$$\frac{a_2b_0 - b_2}{a_0a_2 - 1}, \ \frac{a_3b_1 - b_3}{a_1a_3 - 1}, \ \frac{a_0b_2 - b_0}{a_0a_2 - 1}, \ \frac{a_1b_3 - b_1}{a_1a_3 - 1}, \ \dots,$$

where $a_0a_2 \neq 1$ and $a_1a_3 \neq 1$. Observe that x_0 and x_2 have only even ordered indices and x_1 and x_3 have only odd ordered indices. In addition, x_0 and x_2 have the same denominator and the indices in the numerator shift by 2. We see similar pattern and rhythm in x_1 and x_3 but only with odd ordered indices.

We can see many differences between the periodic cycles in Equation (4.79) with only one periodic sequence $\{a_n\}_{n=0}^{\infty}$ compared to the periodic cycles in Equations (4.83) and (4.84) with two periodic sequences $\{a_n\}_{n=0}^{\infty}$ and $\{b_n\}_{n=0}^{\infty}$. First of all, the periodic cycles are unique with two periodic sequences in Equations (4.83) and (4.84). Second, Equation (4.79) can have a shorter periodic cycle than the autonomous Δ.E.; this is possible with only one periodic sequence. Now from Example 34, the following theorem describes the result when $\{a_n\}_{n=0}^{\infty}$ and $\{b_n\}_{n=0}^{\infty}$ are periodic-4k sequences, $(k \geq 2)$.

Theorem 11. *Suppose that* $\{a_n\}_{n=0}^{\infty}$ *and* $\{b_n\}_{n=0}^{\infty}$ *are periodic sequences with period-4k, $(k \geq 2)$. Then:*

$$x_{n+2} = a_nx_n + b_n, \quad n = 0, 1, \dots,$$

has a unique period-4k cycle if and only if:

$$\prod_{i=1}^{2k} a_{2i-2} \neq 1 \text{ and } \prod_{i=1}^{2k} a_{2i-1} \neq 1,$$

and

$$x_0 = \frac{a_{2k-2}\left[\ \dots\ [a_4[a_2b_0 - b_2] + b_4]\ \dots\ \right] \pm b_{2k-2}}{\left[\prod_{i=1}^{k} a_{2i-2} - 1\right]},$$

$$x_1 = \frac{a_{2k-1}\left[\ \dots\ [a_5[a_3b_1 - b_3] + b_5]\ \dots\ \right] \pm b_{2k-1}}{\left[\prod_{i=1}^{k} a_{2i-1}\right] - 1}.$$

The proof of Theorem 11 will be left as an exercise at the end of the chapter. Moreover, the interesting question is to examine what happens when $\{a_n\}_{n=0}^{\infty}$ and $\{b_n\}_{n=0}^{\infty}$ are of different periods. Similar questions can be addressed in this case as in Open Problems 9 and 10. Further studies on the periodic character can be found in [15, 16, 17], and [18]. Moreover, applications of these periodic structures can be found in **Resonance** and in **Signal Processing in Sigma-Delta Domain** in [14] and [32].

4.8 Third and Higher Order Linear Difference Equations

Here are two examples of third and higher order homogeneous linear difference equations:

(i)   (Third Order) $x_{n+3} - x_{n+2} - 4x_{n+1} + 4x_n = 0$, $n = 0, 1, \ldots$.

(ii) (Fourth Order) $x_{n+4} - 5x_{n+2} + 4x_n = 0$, $n = 0, 1, \ldots$.

From the two examples above, we can inquire a homogeneous linear Δ.E. of order $m \geq 3$ in the form:

$$x_{n+m} + \sum_{i=1}^{m} a_i x_{n+i-1} = 0 \ , \quad n = 0, 1, \ldots , \tag{4.85}$$

where $a_1, a_2, a_3, \ldots, a_m \in \Re$. The periodic character of solutions of Equation (4.85) has been studied by [15, 16, 17], and [18]. These will be the vital questions to address:

(i)   How will the delay $m \geq 3$ affect the periodicity properties?

(ii) How will the relationship between the value of $m \geq 3$ and the periodic sequences $\{a_n\}_{n=0}^{\infty}$ and $\{b_n\}_{n=0}^{\infty}$ affect the periodic traits?

(iii) When is every solution periodic and when do we have unique periodic cycles?

More details on the periodic traits of solutions of Equation (4.85) and conjectures will be conveyed in Chapter 6 to remit these questions.

4.9 Exercises

In problems 1–8, show that the given solution **satisfies** the given Δ.E.

1. $x_n = (2)^n - (-2)^n$ is a solution of $x_{n+2} - 4x_n = 0$.

2. $x_n = 3^{n+1} - 2^{n+1}$ is a solution of $x_{n+2} - 5x_{n+1} + 6x_n = 0$.

3. $x_n = 2 + 2n$ is a solution of $x_{n+2} - 2x_{n+1} + x_n = 0$.

4. $x_n = 3^n + 4^n + 1$ is a solution of $x_{n+2} - 7x_{n+1} + 12x_n = 6$.

5. $x_n = 5(4)^n - 2 - 3n$ is a solution of $x_{n+2} - 5x_{n+1} + 4x_n = 9$.

6. $x_n = 2 + 2^{n+1} - n2^{n-1}$ is a solution of $x_{n+2} - 3x_{n+1} + 2x_n = -2^n$.

7. $x_n = 3^{n+1} + n3^n + \frac{n^2 3^{n-1}}{2}$ is a solution of $x_{n+2} - 6x_{n+1} + 9x_n = 3^n$.

8. $x_n = 1 + 2^{n+1} + (-2)^n$ is a solution of $x_{n+3} - x_{n+2} - 4x_{n+1} + 4x_n = 0$.

In problems 9–17, determine **general solution** to each Δ.E.

9. $x_{n+2} + 4x_{n+1} - 12x_n = 0$, $n = 0, 1, \ldots$.

10. $x_{n+2} - 7x_{n+1} - 8x_n = 0$, $n = 0, 1, \ldots$.

11. $x_{n+2} - 8x_{n+1} + 15x_n = 0$, $n = 0, 1, \ldots$.

12. $x_{n+2} - 8x_{n+1} + 16x_n = 0$, $n = 0, 1, \ldots$.

13. $x_{n+2} + x_{n+1} + x_n = 0$, $n = 0, 1, \ldots$.

14. $x_{n+2} - \sqrt{2}x_{n+1} + x_n = 0$, $n = 0, 1, \ldots$.

15. $4x_{n+2} + x_n = 0$, $n = 0, 1, \ldots$.

16. $x_{n+2} - 2x_{n+1} + 4x_n = 0$, $n = 0, 1, \ldots$.

17. $x_{n+2} + 2x_{n+1} + 2x_n = 0$, $n = 0, 1, \ldots$.

In problems 18–29, determine the **particular solution** to each Δ.E..

18. $x_{n+2} - 6x_{n+1} + 8x_n = 12$, $n = 0, 1, \ldots$.

19. $x_{n+2} - x_{n+1} - 2x_n = 4$, $n = 0, 1, \ldots$.

20. $x_{n+2} + 3x_{n+1} - 4x_n = 6$, $n = 0, 1, \ldots$.

21. $x_{n+2} - 4x_{n+1} + 3x_n = 2$, $n = 0, 1, \ldots$.

22. $x_{n+2} - 5x_{n+1} + 4x_n = 2^n$, $n = 0, 1, \ldots$.

23. $x_{n+2} - 6x_{n+1} + 8x_n = 3^n$, $n = 0, 1, \ldots$.

24. $x_{n+2} - 4x_{n+1} + 5x_n = 5^n$, $n = 0, 1, \ldots$.

25. $x_{n+2} - x_{n+1} - 6x_n = 3^n$, $n = 0, 1, \ldots$.

26. $x_{n+2} + x_{n+1} - 6x_n = (-2)^n + 3^n$, $n = 0, 1, \ldots$.

27. $x_{n+2} - 4x_{n+1} + 3x_n = n$, $n = 0, 1, \ldots$.

28. $x_{n+2} - x_n = n + 2$, $n = 0, 1, \ldots$.

29. $x_{n+2} + x_{n+1} - 2x_n = n + (-2)^n$, $n = 0, 1, \ldots$.

In problems 30–35, solve the following **Initial Value Problem**.

30.
$$
\begin{cases}
x_{n+2} + 3x_{n+1} - 4x_n = 0 , & n = 0, 1, \ldots \\
x_0 = 2, \\
x_1 = 5.
\end{cases}
$$

31.
$$\begin{cases} x_{n+2} - 6x_{n+1} + 8x_n = 0 \ , \quad n = 0,1,\dots . \\ x_0 = 1 , \\ x_1 = 6 . \end{cases}$$

32.
$$\begin{cases} x_{n+2} - 8x_{n+1} + 16x_n = 0 \ , \quad n = 0,1,\dots . \\ x_0 = 0 , \\ x_1 = 1 . \end{cases}$$

33.
$$\begin{cases} x_{n+2} - 7x_{n+1} + 12x_n = 3 \ , \quad n = 0,1,\dots . \\ x_0 = \frac{1}{2} , \\ x_1 = \frac{5}{2} . \end{cases}$$

34.
$$\begin{cases} x_{n+2} - 5x_{n+1} + 4x_n = 3 \ , \quad n = 0,1,\dots . \\ x_0 = 2 , \\ x_1 = 2 . \end{cases}$$

35.
$$\begin{cases} x_{n+2} - 4x_{n+1} + 3x_n = 3^{n+1} \ , \quad n = 0,1,\dots . \\ x_0 = \frac{1}{2} , \\ x_1 = 3 . \end{cases}$$

In problems 36–42, determine **general solution** to each Δ.E.:

36. $x_{n+3} - x_{n+2} + 4x_{n+1} - 4x_n = 0 \ , \quad n = 0,1,\dots .$

37. $x_{n+4} - x_n = 0 \ , \quad n = 0,1,\dots .$

38. $x_{n+4} - x_{n+2} - 2x_n = 0 \ , \quad n = 0,1,\dots .$

39. $x_{n+4} - 25x_{n+2} + 144x_n = 0 \ , \quad n = 0,1,\dots .$

40. $4x_{n+4} + 3x_{n+2} - x_n = 0 \ , \quad n = 0,1,\dots .$

41. $8x_{n+4} - 6x_{n+2} - x_n = 0 \ , \quad n = 0,1,\dots .$

42. $x_{n+3} + x_n = 0 \ , \quad n = 0,1,\dots .$

In problems 43–50, show that every nontrivial solution is periodic and determine the period.

43. $x_{n+2} = x_n - x_{n+1} \ , \quad n = 0,1,\dots .$

44. $x_{n+2} = \sqrt{2}x_{n+1} - x_n \ , \quad n = 0,1,\dots .$

45. $x_{n+2} = \sqrt{3}x_{n+1} - x_n \ , \quad n = 0,1,\dots .$

46. From Exercises 43, 44, and 45, generalize by induction that every nontrivial solution of the Δ.E. is periodic with period-k, $(k \geq 3)$:

$$x_{n+2} = 2cos\left(\frac{2\pi}{k}\right)x_{n+1} - x_n \ , \quad n = 0,1,\dots .$$

In problems 47–67, determine the necessary and sufficient conditions for the existence of periodic solutions:

47. Existence of Periodic Solutions of:

$$x_{n+2} = -(x_{n+1} - x_n) + a_n \quad , \quad n = 0, 1, \dots ,$$

where $\{a_n\}_{n=0}^{\infty}$ is a period-3k sequence, $(k \geq 2)$.

48. Existence of Periodic Solutions of:

$$x_{n+2} = -x_n + a_n \quad , \quad n = 0, 1, \dots ,$$

where $\{a_n\}_{n=0}^{\infty}$ is a period-8 sequence.

49. Existence of Periodic Solutions of:

$$x_{n+2} = -x_n + a_n \quad , \quad n = 0, 1, \dots ,$$

where $\{a_n\}_{n=0}^{\infty}$ is a period-12 sequence.

50. Existence of Periodic Solutions of:

$$x_{n+2} = -x_n + a_n \quad , \quad n = 0, 1, \dots ,$$

where $\{a_n\}_{n=0}^{\infty}$ is a period-4k sequence, $(k \geq 2)$.

51. Existence of Periodic Solutions of:

$$x_{n+2} = -(x_{n+1} + x_n) + a_n \quad , \quad n = 0, 1, \dots ,$$

where $\{a_n\}_{n=0}^{\infty}$ is a period-12 sequence, $(k \geq 2)$.

52. Existence of Periodic Solutions of:

$$x_{n+2} = -(x_{n+1} + x_n) + a_n \quad , \quad n = 0, 1, \dots ,$$

where $\{a_n\}_{n=0}^{\infty}$ is a period-3k sequence, $(k \geq 2)$.

53. Existence of Periodic Solutions of:

$$x_{n+2} = -(x_{n+1} + x_n) + a_n \quad , \quad n = 0, 1, \dots ,$$

where $\{a_n\}_{n=0}^{\infty}$ is a period-9 sequence, $(k \geq 2)$.

54. Existence of Periodic Solutions of:

$$x_{n+2} = (-1)^n x_n + a_n \quad , \quad n = 0, 1, \dots ,$$

where $\{a_n\}_{n=0}^{\infty}$ is a period-2 sequence.

55. Existence of Periodic Solutions of:

$$x_{n+2} = (-1)^n x_n + a_n \quad , \quad n = 0, 1, \ldots ,$$

where $\{a_n\}_{n=0}^{\infty}$ is a period-4 sequence.

56. Existence of Periodic Solutions of:

$$x_{n+2} = (-1)^n x_n + a_n \quad , \quad n = 0, 1, \ldots ,$$

where $\{a_n\}_{n=0}^{\infty}$ is a period-8 sequence.

57. Existence of Periodic Solutions of:

$$x_{n+2} = (-1)^n x_n + a_n \quad , \quad n = 0, 1, \ldots ,$$

where $\{a_n\}_{n=0}^{\infty}$ is a period-12 sequence.

58. Existence of Periodic Solutions of:

$$x_{n+2} = (-1)^n x_n + a_n \quad , \quad n = 0, 1, \ldots ,$$

where $\{a_n\}_{n=0}^{\infty}$ is a period-4k sequence, $(k \in \mathbb{N})$.

59. Existence of Periodic Solutions of:

$$x_{n+2} = \sqrt{2} x_{n+1} - x_n + a_n \quad , \quad n = 0, 1, \ldots ,$$

where $\{a_n\}_{n=0}^{\infty}$ is a period-2 sequence.

60. Existence of Periodic Solutions of:

$$x_{n+2} = \sqrt{2} x_{n+1} - x_n + a_n \quad , \quad n = 0, 1, \ldots ,$$

where $\{a_n\}_{n=0}^{\infty}$ is a period-4 sequence.

61. Existence of Periodic Solutions of:

$$x_{n+2} = \sqrt{3} x_{n+1} - x_n + a_n \quad , \quad n = 0, 1, \ldots ,$$

where $\{a_n\}_{n=0}^{\infty}$ is a period-2 sequence.

62. Existence of Periodic Solutions of:

$$x_{n+2} = \sqrt{3} x_{n+1} - x_n + a_n \quad , \quad n = 0, 1, \ldots ,$$

where $\{a_n\}_{n=0}^{\infty}$ is a period-3 sequence.

63. Existence of Periodic Solutions of:

$$x_{n+2} = \sqrt{3}x_{n+1} - x_n + a_n \quad , \quad n = 0, 1, \dots ,$$

where $\{a_n\}_{n=0}^{\infty}$ is a period-4 sequence.

64. Existence of Periodic Solutions of:

$$x_{n+2} = \sqrt{3}x_{n+1} - x_n + a_n \quad , \quad n = 0, 1, \dots ,$$

where $\{a_n\}_{n=0}^{\infty}$ is a period-6 sequence.

65. Existence of Periodic Solutions of:

$$x_{n+2} = \sqrt{2}x_{n+1} - x_n + a_n \quad , \quad n = 0, 1, \dots ,$$

where $\{a_n\}_{n=0}^{\infty}$ is a period-4 sequence.

66. Existence of Periodic Solutions of:

$$x_{n+2} = x_{n+1} - x_n + a_n \quad , \quad n = 0, 1, \dots ,$$

where $\{a_n\}_{n=0}^{\infty}$ is a period-12 sequence.

67. Existence of Periodic Solutions of:

$$x_{n+2} = x_{n+1} - x_n + a_n \quad , \quad n = 0, 1, \dots ,$$

where $\{a_n\}_{n=0}^{\infty}$ is a period-6k sequence, $(k \geq 2)$.

In problems 68–71, determine the reasons for no existence of periodic solutions:

68. Periodicity of:

$$x_{n+2} = -x_n + a_n \quad , \quad n = 0, 1, \dots ,$$

where $\{a_n\}_{n=0}^{\infty}$ is a period-4 sequence.

69. Periodicity of:

$$x_{n+2} = x_{n+1} - x_n + a_n \quad , \quad n = 0, 1, \dots ,$$

where $\{a_n\}_{n=0}^{\infty}$ is a period-6 sequence.

70. Periodicity of:

$$x_{n+2} = \sqrt{2}x_{n+1} - x_n + a_n \quad , \quad n = 0, 1, \dots ,$$

where $\{a_n\}_{n=0}^{\infty}$ is a period-8 sequence.

71. From Exercises 68, 69, and 70, the Periodicity of the Δ.E.

$$x_{n+2} = 2cos\left(\frac{2\pi}{k}\right)x_{n+1} - x_n + a_n \quad , \quad n = 0, 1, \ldots \, ,$$

where $\{a_n\}_{n=0}^{\infty}$ is a period-k sequence and $\theta = \frac{2\pi}{k}$, $(k \geq 2)$.

In problems 72–76, determine the necessary and sufficient conditions for the existence and uniqueness of periodic solutions.

72. Existence and Uniqueness of Periodic Solutions of:

$$x_{n+2} = a_n x_n + b_n \quad , \quad n = 0, 1, \ldots \, ,$$

where $\{a_n\}_{n=0}^{\infty}$ and $\{b_n\}_{n=0}^{\infty}$ are period-6 sequences, $(k \geq 2)$.

73. Existence and Uniqueness of Periodic Solutions of:

$$x_{n+2} = a_n x_n + b_n \quad , \quad n = 0, 1, \ldots \, ,$$

where $\{a_n\}_{n=0}^{\infty}$ and $\{b_n\}_{n=0}^{\infty}$ are period-2k sequences, $(k \geq 2)$.

74. Existence and Uniqueness of Periodic Solutions of:

$$x_{n+2} = a_n x_n + b_n \quad , \quad n = 0, 1, \ldots \, ,$$

where $\{a_n\}_{n=0}^{\infty}$ is a period-2 sequence and $\{b_n\}_{n=0}^{\infty}$ is a period-4 sequence.

75. Existence and Uniqueness of Periodic Solutions of:

$$x_{n+2} = a_n x_n + b_n \quad , \quad n = 0, 1, \ldots \, ,$$

where $\{a_n\}_{n=0}^{\infty}$ is a period-2 sequence and $\{b_n\}_{n=0}^{\infty}$ is a period-6 sequence.

76. Existence and Uniqueness of Periodic Solutions of:

$$x_{n+2} = a_n x_n + b_n \quad , \quad n = 0, 1, \ldots \, ,$$

where $\{a_n\}_{n=0}^{\infty}$ is a period-2 sequence and $\{b_n\}_{n=0}^{\infty}$ is a period-2l sequence, $(l \geq 2)$.

Problems 80–86 are open-ended questions. Determine if periodic solutions exist. If so, then determine the pattern of the period. If not, then explain why.

81. Periodic Solutions of:

$$x_{n+2} = (-1)^n x_n + C \quad , \quad n = 0, 1, \ldots \, ,$$

where $C \neq 0$.

82. Periodic Solutions of:

$$x_{n+2} = (-1)^n x_{n+1} - x_n \quad , \quad n = 0, 1, \dots .$$

83. Periodic Solutions of:

$$x_{n+2} = (-1)^{n+1} (x_{n+1} + x_n) \quad , \quad n = 0, 1, \dots .$$

84. Periodic Solutions of:

$$x_{n+2} = -a_n (x_{n+1} + x_n) \quad , \quad n = 0, 1, \dots ,$$

where $\{a_n\}_{n=0}^\infty$ is a period-3 sequence.

85. Periodic Solutions of:

$$x_{n+2} = a_n x_{n+1} - b_n x_n \quad , \quad n = 0, 1, \dots ,$$

where $\{a_n\}_{n=0}^\infty$ and $\{b_n\}_{n=0}^\infty$ are period-3 sequences.

86. Periodic Solutions of:

$$x_{n+2} = a_n (x_{n+1} - x_n) + b_n \quad , \quad n = 0, 1, \dots ,$$

where $\{a_n\}_{n=0}^\infty$ and $\{b_n\}_{n=0}^\infty$ are period-3 sequences.

Chapter 5
Periodic Traits of Second Order Nonlinear Difference Equations

5.1 Second Order Nonlinear Difference Equations

Our intent in this chapter is to study the periodicity properties of second order non-linear difference equations. We will focus on examining the periodic traits of rational difference equations and Max-Type difference equations. We will commence with rendering three examples of second order nonlinear difference equations that exhibit periodic solutions and eventually periodic solutions:

(i)   (Special Case of **Rational** Δ.**E.**)

$$x_{n+2} = \frac{Ax_n}{x_{n+1} - 1} \quad , \quad n = 0, 1, \dots .$$

(ii)   (Special Case of **Rational** Δ.**E.**)

$$x_{n+1} = \frac{A + x_n}{\alpha + x_{n+1}} \quad , \quad n = 0, 1, \dots .$$

(iii)   (**Max-Type** Δ.**E.**)

$$x_{n+1} = \max\left\{ \frac{A}{x_n}, \frac{B}{x_{n-1}} \right\}, \, n = 0, 1, \dots$$

The long-term behavior of Rational Difference Equations have been studied by several authors [1, 12, 13, 19, 20, 21, 24, 27, 28, 35]. We will begin with the study of necessary and sufficient conditions for the existence of periodic solutions of Second Order Rational Δ.E. in the form:

$$x_{n+2} = \frac{Ax_{n+1} + Bx_n}{Cx_{n+1} + Dx_n} \quad , \quad n = 0, 1, \dots ,$$

where $A, B, C, D \geq 0$ and the initial conditions $x_0, x_1 \geq 0$.

© Springer Nature Switzerland AG 2018
M. A. Radin, *Periodic Character and Patterns of Recursive Sequences*,
https://doi.org/10.1007/978-3-030-01780-4_5

5.2 Patterns of Periodic Solutions of Second Order Rational Difference Equations

Our aim is to examine the existence and patterns of periodic solutions of Second Order Rational Δ.E. in the form:

$$x_{n+2} = \frac{A + Bx_{n+1} + Cx_n}{D + Ex_{n+1} + Fx_n} \quad , \quad n = 0, 1, \dots , \tag{5.1}$$

where $A, B, C, D, E, F \geq 0$ and the initial conditions $x_0, x_1 \geq 0$. Equation (5.1) has 49 special cases and we will investigate the periodic traits of some of these 49 special cases. The Second Order Riccati Δ.E. is one of the 49 special cases. The next two examples will show the existence and patterns of periodic solutions of the Second Order Riccati Δ.E.

Example 1. Show that every nontrivial solution of:

$$x_{n+2} = \frac{1}{x_n} \quad , \quad n = 0, 1, \dots ,$$

is periodic with period-4 and determine the pattern of the period-4 cycle.

Solution: Notice:

$$x_0 ,$$

$$x_1 ,$$

$$x_2 = \frac{1}{x_0} ,$$

$$x_3 = \frac{1}{x_1} ,$$

$$x_4 = \frac{1}{x_2} = \frac{1}{\left[\frac{1}{x_0}\right]} = x_0 ,$$

$$x_5 = \frac{1}{x_3} = \frac{1}{\left[\frac{1}{x_1}\right]} = x_1 .$$

We then obtain the following period-4 pattern:

$$x_0 , x_1 , \frac{1}{x_0} , \frac{1}{x_1} , \dots .$$

Note that $x_0, x_1 \neq 0$, and the product of four neighboring terms:

$$x_0 \cdot x_1 \cdot \frac{1}{x_0} \cdot \frac{1}{x_1} = 1 .$$

Example 2. Show that every nontrivial solution of:

$$x_{n+2} = \frac{x_n}{x_n - 1} \quad , \quad n = 0, 1, \dots ,$$

is periodic with period-4 and determine the pattern of the period-4 cycle.

Solution: Observe:

$$x_0 ,$$

$$x_1 ,$$

$$x_2 = \frac{x_0}{x_0 - 1} ,$$

$$x_3 = \frac{x_1}{x_1 - 1} ,$$

$$x_4 = \frac{x_2}{x_2 - 1} = \frac{\left[\frac{x_0}{x_0-1}\right]}{\left[\frac{x_0}{x_0-1}\right] - 1} = \frac{x_0}{x_0 - (x_0 - 1)} = x_0 ,$$

$$x_5 = \frac{x_3}{x_3 - 1} = \frac{\left[\frac{x_1}{x_1-1}\right]}{\left[\frac{x_1}{x_1-1}\right] - 1} = \frac{x_1}{x_1 - (x_1 - 1)} = x_1 .$$

Hence we acquire the following period-4 pattern:

$$x_0 , x_1 , \frac{x_0}{x_0 - 1} , \frac{x_1}{x_1 - 1} , \dots .$$

Notice that $x_0, x_1 \neq 1$ and that the product and the sum of all the neighboring terms are always equal:

$$\frac{[x_0 x_1]^2}{(x_0 - 1)(x_1 - 1)} = x_0 + x_1 + \frac{x_0}{x_0 - 1} + \frac{x_1}{x_1 - 1} .$$

The next example will illustrate the existence of period-6 cycles and their patterns.

Example 3. Determine the pattern of periodic cycles of:

$$x_{n+2} = \frac{x_{n+1}}{x_n} \quad , \quad n = 0, 1, \dots .$$

Solution: By iteration we obtain the following period-6 pattern:

$$x_0 , x_1 , \frac{x_1}{x_0} , \frac{1}{x_0} , \frac{1}{x_1} , \frac{x_0}{x_1} , \dots ,$$

where $x_0, x_1 \neq 0$.

The next three examples will be directed on the existence and patterns of period-2 cycles.

Example 4. Determine the necessary and sufficient conditions for the existence of period-2 cycles of:

$$x_{n+2} = A + \frac{x_n}{x_{n+1}} \quad , \quad n = 0, 1, \dots , \tag{5.2}$$

where $A, x_0, x_1 > 0$.

Solution: Set $x_2 = x_0$ and $x_3 = x_1$ and produce the following system of equations:

$$x_0,$$

$$x_1,$$

$$x_2 = A + \frac{x_0}{x_1} = x_0,$$

$$x_3 = A + \frac{x_1}{x_2} = A + \frac{x_1}{x_0} = x_1 ,$$

and we procure the following equality:

$$Ax_1 + x_0 = x_0 x_1 = Ax_0 + x_1 ,$$

that reduces to:

$$A(x_0 - x_1) = x_0 - x_1 ,$$

Thus we see that $A = 1$ as we assumed that $x_0 \neq x_1$. Then Equation (5.2) reduces to:

$$x_{n+2} = 1 + \frac{x_n}{x_{n+1}} \quad , \quad n = 0, 1, \dots .$$

By repeating the process above, we get:

$$x_0,$$

$$x_1,$$

$$x_2 = 1 + \frac{x_0}{x_1} = x_0,$$

$$x_3 = 1 + \frac{x_1}{x_2} = 1 + \frac{x_1}{x_0} = x_1 .$$

We solve for x_1 and get $x_1 = \frac{x_0}{x_0 - 1}$, where $x_0 \neq 1$, and obtain the following period-2 pattern:

$$x_0 , \frac{x_0}{x_0 - 1} , x_0 , \frac{x_0}{x_0 - 1} , \dots .$$

Recall that this identical periodic pattern appeared in the Riccati Δ.E. $x_{n+1} = \frac{x_n}{x_n-1}$, whose periodic solutions are on the hyperbolic curve $y = \frac{x}{x-1}$ (Figure 5.1):

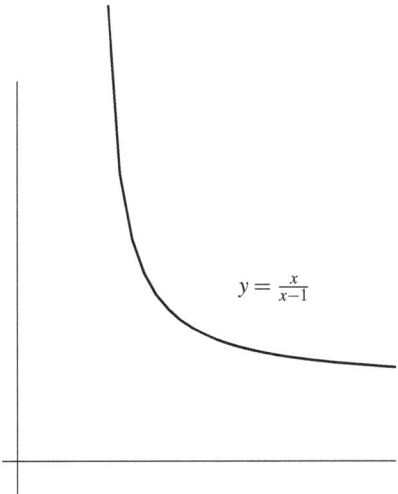

$$y = \frac{x}{x-1}$$

Fig. 5.1 Graph of $y = \frac{x}{x-1}$.

Example 5. Determine the necessary and sufficient conditions for the existence of period-2 cycles of:

$$x_{n+2} = \frac{Ax_n}{1+x_{n+1}} \quad , \quad n = 0, 1, \dots \,, \tag{5.3}$$

where $A > 0$ and $x_0, x_1 \geq 0$.

Solution: Set $x_2 = x_0$ and $x_3 = x_1$ and assemble the following system of equations:

$$x_0,$$

$$x_1,$$

$$x_2 = \frac{Ax_0}{1+x_1} = x_0,$$

$$x_3 = \frac{Ax_1}{1+x_2} = \frac{Ax_1}{1+x_0} = x_1 \,,$$

and procure the following equality:

$$Ax_0 - x_0 = x_0x_1 = Ax_1 - x_1 \,,$$

that reduces to:

$$x_0(A-1) = x_1(A-1) \,.$$

Hence we see that $A = 1$ as we assumed that $x_0 \neq x_1$. Then Equation (5.3) reduces to:

$$x_{n+2} = \frac{x_n}{1 + x_{n+1}} \quad , \quad n = 0, 1, \dots .$$

By repeating the above process, we procure the following system of equations:

$$x_0 ,$$

$$x_1 ,$$

$$x_2 = \frac{x_0}{1 + x_1} = x_0 ,$$

$$x_3 = \frac{x_1}{1 + x_2} = \frac{x_1}{1 + x_0} = x_1 .$$

Therefore $x_0 x_1 = 0$. Then either $x_0 = 0$ or $x_1 = 0$ and acquire one of the following period-2 patterns:

$$0 , x_1 , 0 , x_1 , \dots , \quad \text{or}$$

$$x_0 , 0 , x_0 , 0 , \dots .$$

This is the first time that we see a period-2 pattern where one of the initial conditions must be 0. In this case it is impossible to obtain a positive period-2 cycle.

Example 6. Determine the necessary and sufficient conditions for the existence of period-2 cycles of:

$$x_{n+2} = \frac{A x_n}{1 + x_{n+1} + x_n} \quad , \quad n = 0, 1, \dots , \tag{5.4}$$

where $A > 0$ and $x_0 + x_1 > 0$.

Solution: We set $x_2 = x_0$ and $x_3 = x_1$ and procure the following system of equations:

$$x_0 ,$$

$$x_1 ,$$

$$x_2 = \frac{A x_0}{1 + x_1 + x_0} = x_0 ,$$

$$x_3 = \frac{A x_1}{1 + x_2 + x_1} = \frac{A x_1}{1 + x_0 + x_1} = x_1 ,$$

and acquire the following equality:

$$1 + x_0 + x_1 = A .$$

Hence $A > 1$ as $x_0 + x_1 > 0$. We then obtain one of the following period-2 patterns:

$$x_0 \, , \, A - (1 + x_0) \, , \, x_0 \, , \, A - (1 + x_0) \, , \, \dots \, , \, \text{or}$$

$$A - (1 + x_1) \, , \, x_1 \, , \, A - (1 + x_1) \, , \, x_1 \, , \, \dots \, .$$

In this situation, the periodic solutions are on the line segment $y = A - (1 + x)$ on the restricted interval $[0, A - 1]$ as we can see in the diagram below (Figure 5.2):

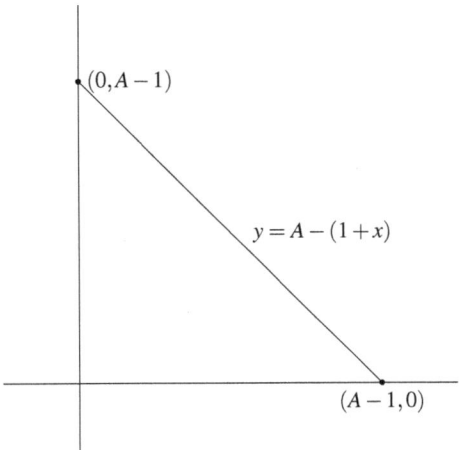

$(0, A - 1)$

$y = A - (1 + x)$

$(A - 1, 0)$

Fig. 5.2 Graph of $y = A - (1 + x)$ on the restricted interval $[0, A - 1]$.

The next section will pursue with determining periodic traits of nonautonomous rational difference equations.

5.3 Periodic Patterns of Second Order Nonautonomous Rational Difference Equations

We will broaden our analysis of periodic essence of solutions (necessary and sufficient conditions for the existence and uniqueness of periodic solutions) of Nonautonomous Rational Difference Equations by introducing a period-k sequence $\{A_n\}_{n=0}^{\infty}$, $(k \geq 2)$. Our plan is to compare the similarities and differences with the autonomous rational difference equations studied in the previous section. We will commence our studies of periodic traits of the following Second Order Nonautonomous Riccati Δ.E.:

$$x_{n+2} = \frac{A_n}{x_n} \, , \quad n = 0, 1, \dots \, , \tag{5.5}$$

where $x_0, x_1 \neq 0$. We will examine several examples and compare the differences between when $\{A_n\}_{n=0}^{\infty}$ is an even ordered sequence and when $\{A_n\}_{n=0}^{\infty}$ is an odd ordered sequence. In the previous chapters we encountered major differences regarding uniqueness of periodic patterns.

Example 7. Suppose that $\{A_n\}_{n=0}^{\infty}$ is a period-2 sequence. Determine the period-4 pattern of:

$$x_{n+2} = \frac{A_n}{x_n} \quad , \quad n = 0, 1, \ldots .$$

Solution: By iteration we procure the following period-4 pattern:

$$x_0 , \ x_1 , \ \frac{A_0}{x_0} , \ \frac{A_1}{x_1} , \ \ldots .$$

The upcoming question to ask: can the Equation (5.5) exhibit period-2 cycles when $\{A_n\}_{n=0}^{\infty}$ is a period-2 sequence? The next example will remit the answer.

Example 8. Suppose that $\{A_n\}_{n=0}^{\infty}$ is a period-2 sequence. Determine the period-2 pattern of:

$$x_{n+2} = \frac{A_n}{x_n} \quad , \quad n = 0, 1, \ldots .$$

Solution: Notice:

$$x_0 ,$$

$$x_1 ,$$

$$x_2 = \frac{A_0}{x_0} = x_0 ,$$

$$x_3 = \frac{A_1}{x_1} = x_1 ,$$

and we acquire:

$$x_0 = \pm\sqrt{A_0} \quad \text{and} \quad x_0 = \pm\sqrt{A_1} ,$$

and hence obtain four distinct period-2 cycles:

- $\sqrt{A_0} , \sqrt{A_1} , \ldots$
- $-\sqrt{A_0} , \sqrt{A_1} , \ldots$
- $\sqrt{A_0} , -\sqrt{A_1} , \ldots$
- $-\sqrt{A_0} , -\sqrt{A_1} . \ldots$

The next question to address: what will happen when $\{A_n\}_{n=0}^{\infty}$ is either an even ordered periodic sequence or an odd ordered periodic sequence? The following examples will convey the answer.

Example 9. Suppose that $\{A_n\}_{n=0}^{\infty}$ is a period-3 sequence. Determine the existence, uniqueness, and pattern of period-3 solutions of:

$$x_{n+2} = \frac{A_n}{x_n} \quad , \quad n = 0, 1, \ldots .$$

Solution: Set $x_6 = x_0$ and $x_7 = x_1$ and we get:

$$x_0 ,$$

$$x_1 ,$$

$$x_2 = \frac{A_0}{x_0} ,$$

$$x_3 = \frac{A_1}{x_1} ,$$

$$x_4 = \frac{A_2}{[x_2]} = \frac{A_2}{\left[\frac{A_0}{x_0}\right]} = \frac{A_2 x_0}{A_0} ,$$

$$x_5 = \frac{A_0}{[x_3]} = \frac{A_0}{\left[\frac{A_1}{x_1}\right]} = \frac{A_0 x_1}{A_1} ,$$

$$x_6 = \frac{A_1}{[x_4]} = \frac{A_1}{\left[\frac{A_2 x_0}{A_0}\right]} = \frac{A_1 A_0}{A_2 x_0} = x_0 ,$$

$$x_7 = \frac{A_2}{[x_5]} = \frac{A_2}{\left[\frac{A_0 x_1}{A_1}\right]} = \frac{A_1 A_2}{A_0 x_1} = x_1 ,$$

and we acquire:

$$x_0 = \pm\sqrt{\frac{A_0 A_1}{A_2}} \text{ and } x_1 = \pm\sqrt{\frac{A_1 A_2}{A_0}} ,$$

Therefore we generate four distinct period-3 cycles:

- $\sqrt{\frac{A_0 A_1}{A_2}}, \sqrt{\frac{A_1 A_2}{A_0}}, \sqrt{\frac{A_2 A_0}{A_1}}, \ldots .$

- $\sqrt{\frac{A_0 A_1}{A_2}}, -\sqrt{\frac{A_1 A_2}{A_0}}, \sqrt{\frac{A_2 A_0}{A_1}}, \ldots .$

- $-\sqrt{\frac{A_0 A_1}{A_2}}, \sqrt{\frac{A_1 A_2}{A_0}}, -\sqrt{\frac{A_2 A_0}{A_1}}, \ldots .$

- $-\sqrt{\frac{A_0 A_1}{A_2}}, -\sqrt{\frac{A_1 A_2}{A_0}}, -\sqrt{\frac{A_2 A_0}{A_1}}, \ldots .$

Observe that from neighbor to neighbor the indices of the sequence $\{A_n\}_{n=0}^{\infty}$ shift by 1.

Example 10. Suppose that $\{A_n\}_{n=0}^{\infty}$ is a period-4 sequence. Determine the existence, uniqueness, and pattern of period-4 solutions of:

$$x_{n+2} = \frac{A_n}{x_n} \quad , \quad n = 0, 1, \dots .$$

Solution: Similar to the previous two examples, set $x_4 = x_0$ and $x_5 = x_1$ and we obtain:

$$x_0 ,$$

$$x_1 ,$$

$$x_2 = \frac{A_0}{x_0} ,$$

$$x_3 = \frac{A_1}{x_1} ,$$

$$x_4 = \frac{A_2}{[x_2]} = \frac{A_2}{\left[\frac{A_0}{x_0}\right]} = \frac{A_2 x_0}{A_0} = x_0 ,$$

$$x_5 = \frac{A_3}{[x_3]} = \frac{A_3}{\left[\frac{A_1}{x_1}\right]} = \frac{A_3 x_1}{A_1} = x_1 .$$

Now observe that $x_4 = x_0$ and $x_5 = x_1$ if and only if

$$A_2 = A_0 \quad \text{and} \quad A_1 = A_3 .$$

Every solution is then periodic with the following period-4 pattern:

$$x_0 , x_1 , \frac{A_0}{x_0} , \frac{A_1}{x_1} , \dots .$$

Example 11. Suppose that $\{A_n\}_{n=0}^{\infty}$ is a period-6 sequence. Determine the existence, uniqueness, and pattern of period-6 solutions of:

$$x_{n+2} = \frac{A_n}{x_n} \quad , \quad n = 0, 1, \dots .$$

Solution: Set $x_6 = x_0$ and $x_7 = x_1$ and we produce:

$$x_0 \, ,$$

$$x_1 \, ,$$

$$x_2 = \frac{A_0}{x_0} \, ,$$

$$x_3 = \frac{A_1}{x_1} \, ,$$

$$x_4 = \frac{A_2}{[x_2]} = \frac{A_2 x_0}{A_0} \, ,$$

$$x_5 = \frac{A_3}{[x_3]} = \frac{A_3 x_1}{A_1} \, ,$$

$$x_6 = \frac{A_4}{[x_4]} = \frac{A_4 A_0}{A_2 x_0} = x_0 \, ,$$

$$x_7 = \frac{A_5}{[x_5]} = \frac{A_5 A_1}{A_3 x_1} = x_1 \, ,$$

and we obtain:

$$x_0 = \pm \sqrt{\frac{A_4 A_0}{A_2}} \quad \text{and} \quad x_1 = \pm \sqrt{\frac{A_5 A_1}{A_3}} \, .$$

Thus we acquire four distinct period-6 cycles:

- $\sqrt{\frac{A_4 A_0}{A_2}} \, , \; \sqrt{\frac{A_5 A_1}{A_3}} \, , \; \sqrt{\frac{A_0 A_2}{A_4}} \, , \; \sqrt{\frac{A_1 A_3}{A_5}} \, , \; \sqrt{\frac{A_2 A_4}{A_0}} \, , \; \sqrt{\frac{A_3 A_5}{A_1}} \, , \; \dots .$

- $-\sqrt{\frac{A_4 A_0}{A_2}} \, , \; \sqrt{\frac{A_5 A_1}{A_3}} \, , \; -\sqrt{\frac{A_0 A_2}{A_4}} \, , \; \sqrt{\frac{A_1 A_3}{A_5}} \, , \; -\sqrt{\frac{A_2 A_4}{A_0}} \, , \; \sqrt{\frac{A_3 A_5}{A_1}} \, , \; \dots .$

- $\sqrt{\frac{A_4 A_0}{A_2}} \, , \; -\sqrt{\frac{A_5 A_1}{A_3}} \, , \; \sqrt{\frac{A_0 A_2}{A_4}} \, , \; -\sqrt{\frac{A_1 A_3}{A_5}} \, , \; \sqrt{\frac{A_2 A_4}{A_0}} \, , \; -\sqrt{\frac{A_3 A_5}{A_1}} \, , \; \dots .$

- $-\sqrt{\frac{A_4 A_0}{A_2}} \, , \; -\sqrt{\frac{A_5 A_1}{A_3}} \, , \; -\sqrt{\frac{A_0 A_2}{A_4}} \, , \; -\sqrt{\frac{A_1 A_3}{A_5}} \, , \; -\sqrt{\frac{A_2 A_4}{A_0}} \, , \; -\sqrt{\frac{A_3 A_5}{A_1}} \, , \; \dots .$

Observe from $x_0 - x_2$ and from $x_2 - x_4$, the indices of the sequence $\{A_n\}_{n=0}^{\infty}$ shift by 2. Similar phenomena occurs from $x_1 - x_3$ and from $x_3 - x_5$.

From Examples 9–11, we can see contrasts of periodic traits of Equation (5.5) when $\{A_n\}_{n=0}^{\infty}$ is periodic with various periods. The following three theorems outline the conclusions.

Theorem 1. *Suppose that* $\{A_n\}_{n=0}^{\infty}$ *is a period-4k sequence, $(k \in \mathbb{N})$. Then every solution of:*

$$x_{n+2} = \frac{A_n}{x_n} \, , \quad n = 0, 1, \dots \, ,$$

is periodic with period-4k if and only if:

$$\prod_{i=1}^{k} A_{4i-4} = \prod_{i=1}^{k} A_{4i-2} \quad \text{and} \quad \prod_{i=1}^{k} A_{4i-3} = \prod_{i=1}^{k} A_{4i-1} .$$

Theorem 2. *Suppose that* $\{A_n\}_{n=0}^{\infty}$ *is a period-*$(4k+2)$ *sequence,* $(k \in \mathbb{N})$. *Then:*

$$x_{n+2} = \frac{A_n}{x_n} \quad , \quad n = 0, 1, \dots ,$$

has four distinct period-$(4k+2)$ *cycles with:*

$$x_0 = \pm \sqrt{\frac{\prod_{i=1}^{k+1} A_{4i-4}}{\prod_{i=1}^{k} A_{4i-2}}} \quad \text{and} \quad x_1 = \pm \sqrt{\frac{\prod_{i=1}^{k+1} A_{4i-3}}{\prod_{i=1}^{k} A_{4i-1}}} .$$

Theorem 3. *Suppose that* $\{A_n\}_{n=0}^{\infty}$ *is a period-*$(2k+1)$ *sequence,* $(k \in \mathbb{N})$. *Then:*

$$x_{n+2} = \frac{A_n}{x_n} \quad , \quad n = 0, 1, \dots ,$$

has four distinct period-$(2k+1)$ *cycles.*

To prove Theorem 3, we require two cases: when $\{A_n\}_{n=0}^{\infty}$ is a period-$(4k-1)$ sequence, $(k \in \mathbb{N})$ and when $\{A_n\}_{n=0}^{\infty}$ is a period-$(4k+1)$ sequence, $(k \in \mathbb{N})$.

Now we will advance with exploring the periodic nature of other nonautonomous rational difference equations. The next series of examples will analyze the periodic make up of solutions of the following Rational Nonautonomous Δ.E.:

$$x_{n+2} = \frac{A_n x_n}{1 + x_{n+1}} \quad , \quad n = 0, 1, \dots , \tag{5.6}$$

where $x_0 + x_1 > 0$ and $\{A_n\}_{n=0}^{\infty}$ is a period-k sequence, $(k \geq 2)$. Our aim is to compare periodic traits of Equation (5.6) with the periodicity of Equation (5.3); differences in the length of the periods depending on whether $\{A_n\}_{n=0}^{\infty}$ is an even period or an odd period and the existence of multiple periodic solutions. We observed similar phenomenon with piece-wise difference equations in Chapter 3.

Example 12. Suppose that $\{A_n\}_{n=0}^{\infty}$ is a period-2 sequence. Determine the necessary and sufficient conditions for the existence of period-2 cycles of:

$$x_{n+2} = \frac{A_n x_n}{1 + x_{n+1}} \quad , \quad n = 0, 1, \dots .$$

Solution: We set $x_2 = x_0$ and $x_3 = x_1$ and we get:

$$x_0,$$

$$x_1,$$

$$x_2 = \frac{A_0 x_0}{1 + x_1} = x_0,$$

$$x_3 = \frac{A_1 x_1}{1 + x_2} = \frac{A_1 x_1}{1 + x_0} = x_1,$$

and obtain one of the following equalities:

$$x_0 (A_0 - 1) = x_0 x_1 \quad \text{or} \quad x_1 (A_1 - 1) = x_0 x_1.$$

Notice that they cannot hold true simultaneously as we assumed that $A_0 \neq A_1$. Compared to Equation (5.3) in Example 5, period-2 solutions will exist if either $A_0 = 1$ or $A_1 = 1$. This leads the analysis to two cases:

- (Case 1:) Suppose that $A_0 = 1$, then similar to Example 5, $x_1 = 0$ with the following period-2 pattern:
$$x_0, 0, x_0, 0, \dots.$$

- (Case 2:) Suppose that $A_1 = 1$, then similar to Example 5, $x_0 = 0$ with the following period-2 pattern:
$$0, x_1, 0, x_1, \dots.$$

In Case 1 and Case 2, the match of parity between A_0 and x_1 and between A_1 and x_0 becomes a vital factor for the existence of period-2 cycles. We did not see such a parity dependence in Equation (5.3).

The next example will assume that $\{A_n\}_{n=0}^{\infty}$ is a period-3 sequence and determine the periodic nature of solutions of Equation (5.6).

Example 13. Suppose that $\{A_n\}_{n=0}^{\infty}$ is a period-3 sequence. Determine the necessary and sufficient conditions for the existence of period-6 cycles of:

$$x_{n+2} = \frac{A_n x_n}{1 + x_{n+1}}, \quad n = 0, 1, \dots.$$

Solution: As in Examples (5) and (12), either $x_0 = 0$ or $x_1 = 0$. In this case let $x_1 = 0$ and set $x_6 = x_0$ and $x_7 = x_1$. The case when $x_0 = 0$ is similar and will be omitted. Notice:

$$x_0,$$

$$x_1 = 0,$$

$$x_2 = \frac{A_0 x_0}{1 + x_1} = A_0 x_0,$$

$$x_3 = \frac{A_1 x_1}{1 + x_2} = \frac{A_1 x_1}{1 + x_0} = 0,$$

$$x_4 = \frac{A_2 \lfloor x_2 \rfloor}{1 + x_3} = A_2 A_0 x_0,$$

$$x_5 = \frac{A_0 x_3}{1 + x_4} = 0,$$

$$x_6 = \frac{A_1 \lfloor x_4 \rfloor}{1 + x_5} = A_1 A_2 A_0 x_0 = x_0,$$

$$x_7 = \frac{A_2 x_5}{1 + x_6} = 0 = x_1.$$

Hence $x_6 = x_0$ if and only if $A_1 A_2 A_0 = 1$ with the following period-6 pattern:

$$x_0, 0, A_0 x_0, 0, A_2 A_0 x_0, 0, A_1 A_2 A_0 x_0, \ldots.$$

Observe that the even ordered indices A_0 and A_2 are multiplied in the beginning of the pattern and the odd ordered index A_1 is multiplied at the very end of the pattern. First of all, it is impossible to obtain a period-2 cycle or a period-3 cycle. Second, period-6 is shortest periodic cycle. In fact, period-6 cycles come from period-2 of Equation (5.3) and the period-3 sequence $\{A_n\}_{n=0}^{\infty}$. On one hand, we can see that $x_1 = x_3 = x_5 = 0$ (all the odd ordered terms are 0). On the other hand, when $x_0 = 0$, we then produce the following period-6 pattern:

$$0, x_1, 0, A_1 x_1, 0, A_0 A_1 x_1, 0, A_2 A_0 A_1 x_1, \ldots,$$

where $x_0 = x_2 = x_4 = 0$ (all the even ordered terms are 0) and the pattern emerges with an odd ordered index A_1 instead.

From Example (13), when $\{A_n\}_{n=0}^{\infty}$ is an odd ordered period-$(2k + 1)$ sequence $(k \in \mathbb{N})$, then we can expect the existence of only period-$2(2k + 1)$ cycles of Equation (5.6). The following theorem summarizes the result.

Theorem 4. *Suppose that $\{A_n\}_{n=0}^{\infty}$ is a period-$(2k + 1)$ sequence, $(k \in \mathbb{N})$. Then every solution of:*

$$x_{n+2} = \frac{A_n x_n}{1 + x_{n+1}}, \quad n = 0, 1, \ldots,$$

is periodic with period-2(2k + 1) if and only if either:

$$x_1 = 0 \text{ and } \prod_{i=1}^{2k+1} A_{i-1} = 1, \text{ or}$$

$$x_0 = 0 \text{ and } \prod_{i=1}^{k+1} A_{i-1} = 1.$$

Proof: *Suppose that $x_1 = 0$. The case when $x_0 = 0$ is similar and will be omitted. Let*

$$P_0 = \prod_{i=1}^{k+1} A_{2i-2} \text{ and } P_1 = \prod_{i=1}^{k} A_{2i-1}.$$

As mentioned in Example 13, the even ordered indices will be multiplied first and then the odd ordered indices. First we will see P_0 and then P_1.

$$x_0,$$

$$x_1 = 0,$$

$$x_2 = \frac{A_0 x_0}{1 + x_1} = A_0 x_0,$$

$$x_3 = 0,$$

$$x_4 = \frac{A_2 [x_2]}{1 + x_3} = A_2 A_0 x_0,$$

$$x_5 = 0,$$

$$x_6 = \frac{A_4 [x_4]}{1 + x_5} = A_4 A_2 A_0 x_0,$$

$$\vdots$$

$$x_{2k+1} = 0,$$

$$x_{2k+2} = P_0 x_0,$$

$$x_{2k+3} = 0,$$

$$x_{2k+4} = A_1 P_0 x_0,$$

$$x_{2k+5} = 0,$$

$$x_{2k+6} = A_3 A_1 P_0 x_0 ,$$

$$\vdots$$

$$x_{4k+1} = 0 ,$$

$$x_{4k+2} = P_1 P_0 x_0 .$$

The result follows.

Now suppose that $\{A_n\}_{n=0}^{\infty}$ is an even ordered period-2k sequence ($k \geq 2$). For the first time we will discern the existence of multiple periodic cycles of Equation (5.6). For instance, in the next example when $\{A_n\}_{n=0}^{\infty}$ is a period-4 sequence, we will discover the existence of period-2 cycles and period-4 cycles.

Example 14. Suppose that $\{A_n\}_{n=0}^{\infty}$ is a period-4 sequence. Determine the necessary and sufficient conditions for the existence of period-2 cycles and period-4 cycles of:

$$x_{n+2} = \frac{A_n x_n}{1 + x_{n+1}} , \quad n = 0, 1, \ldots ,$$

Solution: From Example 12, period-2 solutions will exist if either $A_0 = 1$ or $A_1 = 1$. This time we have four periodic coefficients A_0, A_1, A_2, and A_3. This will guide us into four cases:

- (Case 1:) Suppose that $A_0 = A_2 = 1$, then similar to Example 12, $x_1 = 0$ with the following period-2 pattern:

$$x_0 , 0 , x_0 , 0 , \ldots .$$

- (Case 2:) Suppose that $A_1 = A_3 = 1$, then similar to Example 12, $x_0 = 0$ with the following period-2 pattern:

$$0 , x_1 , 0 , x_1 , \ldots .$$

- (Case 3:) Suppose that $A_0 A_2 = 1$, then similar to Example 12, $x_1 = 0$ with the following period-4 pattern:

$$x_0 , 0 , A_0 x_0 , 0 , \ldots .$$

- (Case 4:) Suppose that $A_1 A_3 = 1$, then similar to Example 12, $x_0 = 0$ with the following period-4 pattern:

$$0 , x_1 , 0 , A_1 x_1 , \ldots .$$

From Example 14, when $\{A_n\}_{n=0}^{\infty}$ is an even ordered period we can then expect the existence of multiple even ordered periodic cycles only. In the next example, we will assume that $\{A_n\}_{n=0}^{\infty}$ is a period-8 sequence and discover periodic cycles with period-2, period-4, and period-8. One of the initial conditions must be 0 for the existence of any periodic cycles of Equation (5.6) as we experienced in previous examples.

Example 15. Suppose that $\{A_n\}_{n=0}^{\infty}$ is a period-8 sequence. Determine the necessary and sufficient conditions for the existence of the period-2 cycles, period-4 cycles, and period-8 cycles of:

$$x_{n+2} = \frac{A_n x_n}{1 + x_{n+1}} \quad , \quad n = 0, 1, \ldots .$$

Solution: We will begin with the existence of period-2 cycles. Similar to the previous examples, period-2 cycles exist if either:

- $x_0 = 0$ and $A_0 = A_2 = A_4 = A_6 = 1$, or

- $x_1 = 0$ and $A_1 = A_3 = A_5 = A_7 = 1$.

Furthermore, period-4 cycles exist if either:

- $x_0 = 0$ and $A_0 A_4 = 1$, or

- $x_1 = 0$ and $A_1 A_5 = 1$.

Moreover, period-8 cycles exist if either:

- $x_0 = 0$ and $A_0 A_4 A_6 A_8 = 1$, or

- $x_1 = 0$ and $A_1 A_3 A_5 A_7 = 1$.

From Examples (12)–(15), the following theorem recapitulates the results.

Theorem 5. *Suppose that $\{A_n\}_{n=0}^{\infty}$ is a period-4k sequence, $(k \geq 2)$. Then:*

$$x_{n+2} = \frac{A_n x_n}{1 + x_{n+1}} \quad , \quad n = 0, 1, \ldots ,$$

is periodic with:

(1) Period-2 if for all $i = 1, 2, \ldots, 2k$ either:

$$x_1 = 0 \text{ and } A_{2i-2} = 1, \text{ or } x_0 = 0 \text{ and } A_{2i-1} = 1 .$$

(2) Period-4 if either:

$$x_1 = 0 \text{ and } A_0 A_2 = 1, \text{ or } x_0 = 0 \text{ and } A_1 A_3 = 1 .$$

(3) *Period-2k if either:*

$$x_1 = 0 \quad \text{and} \quad \prod_{i=1}^{\frac{k}{2}} A_{2i-2} = 1 \ , \quad \text{or}$$

$$x_0 = 0 \quad \text{and} \quad \prod_{i=1}^{\frac{k}{2}} A_{2i-1} = 1 \ .$$

(4) *Period-4k if either:*

$$x_1 = 0 \quad \text{and} \quad \prod_{i=1}^{2k} A_{2i-2} = 1 \ , \quad \text{or}$$

$$x_0 = 0 \quad \text{and} \quad \prod_{i=1}^{2k} A_{2i-1} = 1 \ .$$

From Theorem 5 we will pose the following conjecture:

Conjecture 1. Suppose that $\{A_n\}_{n=0}^\infty$ is a period-2k sequence, $(k \geq 2)$. Then:

$$x_{n+2} = \frac{A_n x_n}{1 + x_{n+1}} \ , \quad n = 0, 1, \dots ,$$

has even ordered period-p solutions such that $2k = Np$ for some $N \in \mathbb{N}$.

The next series of examples will analyze the periodicity of:

$$x_{n+2} = \frac{A_n x_n}{1 + x_{n+1} + x_n} \ , \quad n = 0, 1, \dots , \tag{5.7}$$

where $x_0 + x_1 > 0$, and $\{A_n\}_{n=0}^\infty$ is a period-k sequence, $(k \geq 2)$. Similar phenomenon will occur with the existence of multiple periodic cycles and the differences in periodicity properties depending on whether $\{A_n\}_{n=0}^\infty$ is an even ordered period or an odd ordered period. However, the primary contrast with Equation (5.4) will be the uniqueness of periodic cycles.

Example 16. Suppose that $\{A_n\}_{n=0}^\infty$ is a period-2 sequence. Determine the necessary and sufficient conditions for the existence of period-2 cycles of:

$$x_{n+2} = \frac{A_n x_n}{1 + x_{n+1} + x_n} \ , \quad n = 0, 1, \dots .$$

Solution: Set $x_2 = x_0$ and $x_3 = x_1$ and we get:

$$x_0 \, ,$$

$$x_1 \, ,$$

$$x_2 = \frac{A_0 x_0}{1 + x_1 + x_0} = x_0 \, ,$$

$$x_3 = \frac{A_1 x_1}{1 + x_2 + x_1} = \frac{A_1 x_1}{1 + x_0 + x_1} = x_1 \, ,$$

and obtain two equalities:

$$A_0 = 1 + x_0 + x_1 \, , \text{ or}$$

$$A_1 = 1 + x_0 + x_1 \, .$$

Notice that they cannot hold true simultaneously as we assumed that $A_0 \neq A_1$. This steers us into two cases:

- (Case 1:) Suppose that $A_0 = 1 + x_0 + x_1$. Then similar to Example 6, let $x_0 = A_0 - (x_1 + 1)$ and we get:

$$x_0 = A_0 - (x_1 + 1),$$

$$x_1,$$

$$x_2 = \frac{A_0 x_0}{1 + x_1 + [x_0]} = \frac{A_0 x_0}{1 + x_1 + [A_0 - (x_1 + 1)]} = x_0,$$

$$x_3 = \frac{A_1 x_1}{1 + x_2 + x_1} = \frac{A_1 x_1}{1 + [x_0] + x_1}$$

$$= \frac{A_1 x_1}{1 + [A_0 - (x_1 + 1)] + x_1} = \frac{A_1 x_1}{A_0} \, .$$

Note that $x_3 = x_1$ if and only if $x_1 = 0$ as we assumed that $A_0 \neq A_1$ and acquire the following unique period-2 pattern:

$$A_0 - 1 \, , \, 0 \, , \, A_0 - 1 \, , \, 0 \, , \, \dots \, .$$

- (Case 2:) Suppose that $A_0 = 1 + x_0 + x_1$, then similar to Case 1, we obtain the following unique period-2 pattern:

$$0 \, , \, A_1 - 1 \, , \, 0 \, , \, A_1 - 1 \, , \, \dots \, .$$

In Example 12, the parity match becomes an essential factor for the existence of the period-2 cycle in Case 1 and in Case 2. Furthermore, it is of paramount importance to note that period-2 solutions of Equation (5.4) in Example 5.4 were on a line segment $y = A - (x + 1)$. Contrary, Equation (5.7) has two unique period-2 cycles instead.

The next example will assume that $\{A_n\}_{n=0}^{\infty}$ is a period-3 sequence and determine the periodic nature of solutions of Equation (5.7).

Example 17. Suppose that $\{A_n\}_{n=0}^{\infty}$ is a period-3 sequence. Determine the necessary and sufficient conditions for the existence of period-6 cycles of:

$$x_{n+2} = \frac{A_n x_n}{1 + x_{n+1} + x_n} \quad , \quad n = 0, 1, \dots .$$

Solution: Similar to the previous example, we will analyze two cases when $x_0 = 0$ and when $x_1 = 0$. Suppose that $x_1 = 0$. Then:

$$x_0 ,$$

$$x_1 = 0 ,$$

$$x_2 = \frac{A_0 x_0}{1 + x_1 + x_0} = \frac{A_0 x_0}{1 + x_0} ,$$

$$x_3 = 0 ,$$

$$x_4 = \frac{A_2 x_0}{1 + x_3 + x_2} = \frac{A_2 x_0}{1 + x_2} = \frac{A_0 A_2 x_0}{1 + x_0 + A_0 x_0} ,$$

$$x_5 = 0 ,$$

$$x_6 = \frac{A_1 x_4}{1 + x_5 + x_4} = \frac{A_1 x_4}{1 + x_4} = \frac{A_1 A_0 A_2 x_0}{1 + x_0 + A_0 x_0 + A_0 A_2 x_0} = x_0 .$$

Therefore:

$$x_0 = \frac{A_0 A_1 A_2 - 1}{1 + A_0 + A_0 A_2} .$$

Similarly, set $x_0 = 0$ and we obtain:

$$x_1 = \frac{A_0 A_1 A_2 - 1}{1 + A_1 + A_1 A_0} .$$

The two unique period-6 cycles exist if and only if $A_0 A_1 A_2 \neq 1$.

From Example (17) we can conclude that when $\{A_n\}_{n=0}^{\infty}$ is an odd ordered period-$(2k + 1)$, then we can expect the existence of only period-$2(2k + 1)$ cycles, $(k \in \mathbb{N})$. The following theorem summarizes the results and the proof will be left as an exercise.

Theorem 6. *Suppose that* $\{A_n\}_{n=0}^{\infty}$ *is a period-$(2k+1)$ sequence, $(k \in \mathbb{N})$. Then every solution of:*

$$x_{n+2} = \frac{A_n x_n}{1 + x_{n+1} + x_n} \quad , \quad n = 0, 1, \ldots ,$$

is periodic with period-$2(2k+1)$ if and only if $\left[\prod_{i=1}^{2k+1} A_{i-1}\right] \neq 1$ and either:

$$x_1 = 0 \text{ and } x_0 = \frac{\left[\prod_{i=1}^{2k+1} A_{i-1}\right] - 1}{1 + A_0 + A_0 A_2 + \ldots + \prod_{i=1}^{k+1} A_{2i-2}} , \quad \text{or}$$

$$x_0 = 0 \text{ and } x_1 = \frac{\left[\prod_{i=1}^{2k+1} A_{i-1}\right] - 1}{1 + A_1 + A_1 A_3 + \ldots + A_0 \left[\prod_{i=1}^{k} A_{2i-1}\right]} .$$

Now we will continue with more examples that analyze the periodic traits of Equation (5.7) when $\{A_n\}_{n=0}^{\infty}$ is an even ordered period. The existence of multiple periodic cycles will occur as we experienced with Equation (5.6).

Example 18. Suppose that $\{A_n\}_{n=0}^{\infty}$ is a period-4 sequence. Determine the necessary and sufficient conditions for the existence of period-2 cycles and period-4 cycles of:

$$x_{n+2} = \frac{A_n x_n}{1 + x_{n+1} + x_n} \quad , \quad n = 0, 1, \ldots .$$

Solution: We will break up into two cases when $x_0 = 0$ and when $x_1 = 0$. First suppose $x_1 = 0$ and:

$$x_0 ,$$

$$x_1 = 0 ,$$

$$x_2 = \frac{A_0 x_0}{1 + x_1 + x_0} = \frac{A_0 x_0}{1 + x_0} ,$$

$$x_3 = 0 ,$$

$$x_4 = \frac{A_2 x_0}{1 + x_3 + x_2} = \frac{A_2 x_0}{1 + x_2} = \frac{A_0 A_2 x_0}{1 + x_0 + A_0 x_0} = x_0 .$$

Hence:

$$x_0 = \frac{A_0 A_2 - 1}{1 + A_0} .$$

Similarly, set $x_0 = 0$ and obtain:

$$x_1 = \frac{A_1 A_3 - 1}{1 + A_1} .$$

The two unique period-4 cycles exist if and only if $A_0 A_2 \neq 1$ and $A_1 A_3 \neq 1$. Now we will examine the necessary and sufficient condition for the existence of period-2 cycles. Similar to Example 16, suppose that $x_0 = 0$ and $x_1 = A_1 - 1$, then:

$$x_0 = 0 ,$$

$$x_1 = A_1 - 1 ,$$

$$x_2 = 0 ,$$

$$x_3 = \frac{A_1 x_1}{1 + x_2 + x_1} = A_1 - 1 ,$$

$$x_4 = 0 ,$$

$$x_5 = \frac{A_3 x_3}{1 + x_4 + x_3} = \frac{A_3}{A_1} [A_1 - 1] .$$

Thus the period-2 cycle will exist if either:

- $x_0 = 0$, $x_1 = A_1 - 1$ and $A_1 = A_3$, or

- $x_1 = 0$, $x_0 = A_0 - 1$ and $A_0 = A_2$.

From this example we can conclude that when $\{A_n\}_{n=0}^{\infty}$ is an even ordered period, the existence of multiple even ordered periodic cycles occurs only. In the next example, we will assume that $\{A_n\}_{n=0}^{\infty}$ is a period-8 sequence and we can expect periodic solutions with period-2, period-4, and period-8. Recall that one of the initial conditions must be 0 for any periodic cycles to exist as we observed in the previous examples.

Example 19. Suppose that $\{A_n\}_{n=0}^{\infty}$ is a period-8 sequence. Determine the necessary and sufficient conditions for the existence of period-2 cycles, period-4 cycles, and period-8 cycles of:

$$x_{n+2} = \frac{A_n x_n}{1 + x_{n+1} + x_n} , \quad n = 0, 1, \ldots .$$

Solution: Similar to the previous examples, period-2 cycle exists if either:

- $x_0 = 0$, $x_1 = A_1 - 1$ and $A_1 = A_3 = A_5 = A_7$, or

- $x_1 = 0$, $x_0 = A_0 - 1$ and $A_0 = A_2 = A_4 = A_6$.

Furthermore, period-4 cycle exists if either:

- $x_0 = 0$, $x_1 = \frac{A_1 A_3 - 1}{1 + A_1}$ and $A_1 = A_5$ and $A_3 = A_7$, or

- $x_1 = 0$, $x_0 = \frac{A_0 A_2 - 1}{1 + A_0}$ and $A_0 = A_4$ and $A_2 = A_6$.

Moreover, period-8 cycle exists if either:

- $x_0 = 0$ and $x_1 = \frac{A_1 A_3 A_5 A_7 - 1}{1 + A_1 + A_1 A_3 + A_1 A_3 A_5}$, or

- $x_1 = 0$ and $x_0 = \frac{A_0 A_2 A_4 A_6 - 1}{1 + A_0 + A_0 A_2 + A_0 A_2 A_4}$.

Therefore from Example 16, Example 18, and Example 19, the following theorem describes the results. The proof of the theorem will be left as an exercise at the end of the chapter.

Theorem 7. *Suppose that* $\{A_n\}_{n=0}^{\infty}$ *is a period-4k sequence,* $(k \geq 2)$. *Then:*

$$x_{n+2} = \frac{A_n x_n}{1 + x_{n+1} + x_n} \quad , \quad n = 0, 1, \ldots ,$$

is periodic with:

- *Period-2 if for all* $i = 1, 2, \ldots, 2k - 1$ *either:*

$$x_0 = 0 \ , \ x_1 = A_1 - 1 \text{ and } A_{2i-1} = A_{2i+1} \ , \ \text{ or}$$
$$x_1 = 0 \ , \ x_0 = A_0 - 1 \text{ and } A_{2i-2} = A_{2i} .$$

- *Period-4 if for all* $i = 1, 2, \ldots, 2k - 1$, *either:*

(i)

$$A_{4i-3} = A_{4i+1} \ , \ A_{4i-1} = A_{4i+3} \ , \ \text{and}$$

$$x_0 = 0 \text{ and } x_1 = \frac{A_1 A_3 - 1}{1 + A_1} .$$

(ii)

$$A_{4i-4} = A_{4i} \ , \ A_{4i-2} = A_{4i+2} \ , \ \text{and}$$

$$x_1 = 0 \text{ and } x_0 = \frac{A_0 A_2 - 1}{1 + A_0} .$$

- *Period-2k if either:*

(i)

$$A_1 = A_{2k+1} \ , \ A_3 = A_{2k+3} \ , \ \text{and}$$

$$x_0 = 0 \text{ and } x_1 = \frac{\left[\prod_{i=1}^{k} A_{2i-1}\right] - 1}{1 + A_1 + A_1 A_3 + \ldots + \left[\prod_{i=1}^{k-1} A_{2i-1}\right]} .$$

(ii)

$$A_0 = A_{2k} \ , \ A_2 = A_{2k+2} \ , \text{and}$$

$$x_1 = 0 \text{ and } x_0 = \frac{\left[\prod_{i=1}^{k} A_{2i-2}\right] - 1}{1 + A_0 + A_0 A_2 + \ldots + \left[\prod_{i=1}^{k-1} A_{2i-2}\right]} .$$

- *Period-4k if either:*

$$x_0 = 0 \quad \text{and} \quad x_1 = \frac{\left[\prod_{i=1}^{2k} A_{2i-1}\right] - 1}{1 + A_1 + A_1 A_3 + \ldots + \left[\prod_{i=1}^{2k-1} A_{2i-1}\right]} \, , \text{ or}$$

$$x_1 = 0 \quad \text{and} \quad x_0 = \frac{\left[\prod_{i=1}^{2k} A_{2i-2}\right] - 1}{1 + A_0 + A_0 A_2 + \ldots + \left[\prod_{i=1}^{2k-1} A_{2i-2}\right]} \, .$$

From Theorem 7 we will pose the following conjecture:

Conjecture 2. Suppose that $\{A_n\}_{n=0}^{\infty}$ is a period-2k sequence, $(k \geq 2)$. Then:

$$x_{n+2} = \frac{A_n x_n}{1 + x_{n+1} + x_n} \, , \quad n = 0, 1, \ldots \, ,$$

has even ordered period-p cycles provided that $2k = Np$ for some $N \in \mathbb{N}$.

5.4 Periodic and Eventually Periodic Solutions of Max-Type Difference Equations

In this section we will examine the existence and patterns of periodic and eventually periodic solutions of the following Max-Type Δ.E.:

$$x_{n+2} = \max\left\{\frac{A}{x_{n+1}}, \frac{B}{x_n}\right\}, \quad n = 0, 1, \ldots \, ,$$

where $A, B, x_0, x_1 > 0$. By a change of variables, the above Δ.E. can be rewritten as:

$$x_{n+2} = \max\left\{\frac{1}{x_{n+1}}, \frac{C}{x_n}\right\}, \quad n = 0, 1, \ldots \, , \tag{5.8}$$

where $C, x_0, x_1 > 0$. In [2], it was shown that every positive solution of Equation (5.8) is eventually periodic with the following periods:

$$\begin{cases} 2 \text{ if } C < 1 \, , \\ 3 \text{ if } C = 1 \, , \\ 4 \text{ if } C > 1 \, . \end{cases}$$

First of all, the left-hand side of Equation (5.8) is the Δ.E.:

$$x_{n+2} = \frac{1}{x_{n+1}} \, ,$$

which is the Riccati Δ.E. with period-2 solutions. Second, the right-hand side of Equation (5.8) is the Δ.E.:

$$x_{n+2} = \frac{C}{x_n},$$

which is the **Delayed Riccati Δ.E.** with period-4 solutions. We will study the case when $C < 1$ and show that every positive solution of Equation (5.8) is either periodic with period-2 or eventually periodic with period-2 and describe the pattern of the transient terms. Other cases are similar and will left as exercises at the end of the chapter. Equation (5.8) will not exhibit unique periodic solutions compared with piece-wise difference equations. The periodic essence of Max-Type Difference Equations was investigated by several authors in [2, 7, 8, 9, 25, 26, 30, 27, 28, 35].

Theorem 8. *Let* $\{x_n\}_{n=0}^{\infty}$ *be a solution of:*

$$x_{n+2} = max\left\{\frac{1}{x_{n+1}}, \frac{C}{x_n}\right\}, \quad n = 0, 1, \ldots,$$

where $x_0, x_1 > 0$, *and* $0 < C < 1$. *There exists* $N \geq 0$ *such that*

$$x_{N+1} = \frac{1}{x_N}.$$

Proof: *For the sake of contradiction, suppose that* $x_{n+2} = \frac{C}{x_n}$ *for all* $n \geq 0$. *Then we obtain:*

$$x_0,$$

$$x_1,$$

$$x_2 = max\left\{\frac{1}{x_1}, \frac{C}{x_0}\right\} = \frac{C}{x_0} \quad \left(\text{if } \frac{x_0}{x_1} < C < 1\right),$$

$$x_3 = max\left\{\frac{1}{[x_2]}, \frac{C}{x_1}\right\} = max\left\{\frac{1}{\left[\frac{C}{x_0}\right]}, \frac{C}{x_1}\right\} = max\left\{\frac{x_0}{C}, \frac{C}{x_1}\right\}$$

$$= \frac{C}{x_1} \quad \left(\text{if } x_0 x_1 < C^2 < 1\right),$$

$$x_4 = max\left\{\frac{1}{[x_3]}, \frac{C}{[x_2]}\right\} = max\left\{\frac{1}{\left[\frac{C}{x_1}\right]}, \frac{C}{\left[\frac{C}{x_0}\right]}\right\} = max\left\{\frac{x_1}{C}, x_0\right\}$$

$$= x_0 \quad \left(\text{if } \frac{x_1}{x_0} < C < 1\right).$$

Then we procure the following two inequalities:

$$\frac{x_0}{x_1} < C \quad, \quad \text{and}$$

$$\frac{x_1}{x_0} < C,$$

which is clearly a contradiction. Thus Equation (5.8) cannot have period-4 solutions when $C < 1$. Therefore, when $C < 1$, it suffices to consider the following initial conditions:

$$x_0 \quad \text{and} \quad x_1 = \frac{1}{x_0}.$$

The following example will outline the necessary and sufficient conditions for every solution of Equation (5.8) to be periodic with period-2.

Example 20. Suppose that $C < 1$. Determine the necessary and sufficient conditions for the existence of period-2 solutions of:

$$x_{n+2} = \max\left\{ \frac{1}{x_{n+1}}, \frac{C}{x_n} \right\} \quad, \quad n = 0, 1, \dots .$$

Solution: From Theorem 8 assume that $x_1 = \frac{1}{x_0}$. Then:

$$x_0 ,$$

$$x_1 = \frac{1}{x_0} ,$$

$$x_2 = \max\left\{ \frac{1}{[x_1]}, \frac{C}{x_0} \right\} = \max\left\{ \frac{1}{\left[\frac{1}{x_0}\right]}, \frac{C}{x_0} \right\} = \max\left\{ x_0, \frac{C}{x_0} \right\}$$

$$= x_0 \quad (\text{if } x_0^2 > C) ,$$

$$x_3 = \max\left\{ \frac{1}{[x_2]}, \frac{C}{[x_1]} \right\} = \max\left\{ \frac{1}{x_0}, \frac{C}{\left[\frac{1}{x_0}\right]} \right\} = \max\left\{ \frac{1}{x_0}, C x_0 \right\}$$

$$= \frac{1}{x_0} \quad \left(\text{if } x_0^2 < \frac{1}{C} \right) .$$

We obtain a period-2 cycle if and only if:

$$C < x_0^2 < \frac{1}{C} .$$

Hence if either $x_0^2 < C$ or if $x_0^2 > \frac{1}{C}$, then Equation (5.8) will have eventually periodic solutions with transient terms. Now the fundamental question to ask: exactly how many transient terms will Equation (5.8) have and under what criteria? The next two examples will address the answers to these two questions. In fact, Equation (5.8) will portray 3k transient terms, $(k \in \mathbb{N})$.

Example 21. Suppose that $C < 1$ and $x_0^2 < C$. Determine the necessary and sufficient conditions for the existence of eventually period-2 solutions with three transient terms of:

$$x_{n+2} = \max\left\{\frac{1}{x_{n+1}}, \frac{C}{x_n}\right\} , \quad n = 0, 1, \dots .$$

Solution: As in Example 20, let $x_1 = \frac{1}{x_0}$. Then:

x_0 ,

$$x_1 = \frac{1}{x_0} ,$$

$$x_2 = \max\left\{\frac{1}{[x_1]}, \frac{C}{x_0}\right\} = \max\left\{\frac{1}{\left[\frac{1}{x_0}\right]}, \frac{C}{x_0}\right\} = \max\left\{x_0, \frac{C}{x_0}\right\}$$

$$= \frac{C}{x_0} \ \left(\text{as } x_0^2 < C\right) ,$$

$$x_3 = \max\left\{\frac{1}{[x_2]}, \frac{C}{[x_1]}\right\} = \max\left\{\frac{1}{\left[\frac{C}{x_0}\right]}, \frac{C}{\left[\frac{1}{x_0}\right]}\right\} = \max\left\{\frac{x_0}{C}, Cx_0\right\}$$

$$= \frac{x_0}{C} \ (\text{as } C < 1) ,$$

$$x_4 = \max\left\{\frac{1}{[x_3]}, \frac{C}{[x_2]}\right\} = \max\left\{\frac{1}{\left[\frac{x_0}{C}\right]}, \frac{C}{\left[\frac{C}{x_0}\right]}\right\} = \max\left\{\frac{C}{x_0}, x_0\right\}$$

$$= \frac{C}{x_0} \ \left(\text{as } x_0^2 < C\right) ,$$

$$x_5 = \max\left\{\frac{1}{[x_4]}, \frac{C}{[x_3]}\right\} = \max\left\{\frac{1}{\left[\frac{C}{x_0}\right]}, \frac{C}{\left[\frac{x_0}{C}\right]}\right\} = \max\left\{\frac{x_0}{C}, \frac{C^2}{x_0}\right\}$$

$$= \frac{x_0}{C} = x_3 \ \left(\text{if } C^3 < x_0^2 < C\right) ,$$

$$x_6 = \max\left\{\frac{1}{[x_5]}, \frac{C}{[x_4]}\right\} = \max\left\{\frac{1}{\left[\frac{x_0}{C}\right]}, \frac{C}{\left[\frac{C}{x_0}\right]}\right\} = \max\left\{\frac{C}{x_0}, x_0\right\}$$

$$= \frac{C}{x_0} = x_4 \ \left(\text{as } x_0^2 < C\right) .$$

Thus when:

$$C^3 < x_0^2 < C,$$

we obtain three transient terms in square brackets prior to the period-2 pattern:

$$\left[\mathbf{x_0} , \frac{\mathbf{1}}{\mathbf{x_0}} , \frac{\mathbf{C}}{\mathbf{x_0}} \right] , \frac{x_0}{C} , \frac{C}{x_0} , \dots .$$

Example 22. Suppose that $C < 1$ and $x_0^2 < C^3$. Determine the necessary and sufficient conditions for the existence of eventually period-2 solutions with six transient terms of:

$$x_{n+2} = \max \left\{ \frac{1}{x_{n+1}} , \frac{C}{x_n} \right\} , \quad n = 0, 1, \dots .$$

Solution: Let $x_1 = \frac{1}{x_0}$. Then:

$$x_0 ,$$

$$x_1 = \frac{1}{x_0} ,$$

$$x_2 = \max \left\{ \frac{1}{[x_1]} , \frac{C}{x_0} \right\} = \max \left\{ x_0, \frac{C}{x_0} \right\} = \frac{C}{x_0} ,$$
$$\text{(as } x_0^2 < C) ,$$

$$x_3 = \max \left\{ \frac{1}{[x_2]} , \frac{C}{[x_1]} \right\} = \max \left\{ \frac{x_0}{C} , Cx_0 \right\} = \frac{x_0}{C} ,$$
$$\text{(as } C < 1) ,$$

$$x_4 = \max \left\{ \frac{1}{[x_3]} , \frac{C}{[x_2]} \right\} = \max \left\{ \frac{C}{x_0} , x_0 \right\} = \frac{C}{x_0} ,$$
$$\text{(as } x_0^2 < C) ,$$

$$x_5 = \max \left\{ \frac{1}{[x_4]} , \frac{C}{[x_3]} \right\} = \max \left\{ \frac{x_0}{C} , \frac{C^2}{x_0} \right\} = \frac{C^2}{x_0} ,$$
$$\text{(as } x_0^2 < C^3) ,$$

$$x_6 = \max \left\{ \frac{1}{[x_5]} , \frac{C}{[x_4]} \right\} = \max \left\{ \frac{x_0}{C^2} , x_0 \right\} = \frac{x_0}{C^2} ,$$
$$\text{(as } C < 1) ,$$

$$x_7 = \max \left\{ \frac{1}{[x_6]} , \frac{C}{[x_5]} \right\} = \max \left\{ \frac{C^2}{x_0} , \frac{x_0}{C} \right\} = \frac{C^2}{x_0} ,$$
$$\text{(as } x_0^2 < C^3) ,$$

$$x_8 = \max \left\{ \frac{1}{[x_7]} , \frac{C}{[x_6]} \right\} = \max \left\{ \frac{x_0}{C^2} , \frac{C^3}{x_0} \right\} = \frac{x_0}{C^2} = x_6 .$$
$$\left(\text{if } C^5 < x_0^2 < C^3 \right) .$$

Hence when:

$$C^5 < x_0^2 < C^3,$$

we acquire six transient in square brackets prior to the period-2 pattern:

$$\left[x_0 , \frac{1}{x_0} , \frac{C}{x_0} , \frac{x_0}{C} , \frac{C}{x_0} , \frac{C^2}{x_0} \right] , \frac{x_0}{C^2} , \frac{C^2}{x_0} , \ldots .$$

It is interesting to analyze that the six transient terms can be decomposed into three groups of geometric sequences.

From Examples 21 and 22, there exists $k \in \mathbb{N}$ such that:

$$C^{2k+1} < x_0^2 < C^{2k-1} ,$$

as shown in the diagram below (Figure 5.3).

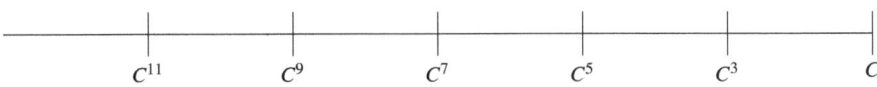

$$C^{11} \qquad C^9 \qquad C^7 \qquad C^5 \qquad C^3 \qquad C$$

Fig. 5.3 We choose x_0^2 in one of the subintervals $\left[C^{2k+1} , C^{2k-1} \right]$ for some $k \in \mathbb{N}$.

Furthermore, from Examples 21 and 22, the following theorem describes the results.

Theorem 9. *Every solution of Equation (5.8) is eventually periodic with period-2 with 3k transient terms if:*

(1) $C < 1$

(2) *There exists $k \in \mathbb{N}$ such that:*

$$C^{2k+1} < x_0^2 < C^{2k-1} .$$

Furthermore, when $x_0^2 > C$, we can similarly conclude that there exists $k \in \mathbb{N}$ such that:

$$\frac{1}{C^{2k-1}} < x_0^2 < \frac{1}{C^{2k+1}} ,$$

as demonstrated in the diagram below (Figure 5.4):

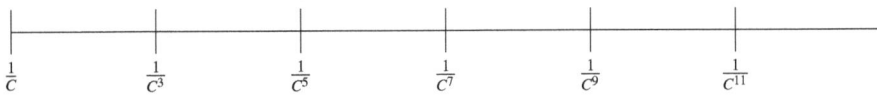

$$\frac{1}{C} \qquad \frac{1}{C^3} \qquad \frac{1}{C^5} \qquad \frac{1}{C^7} \qquad \frac{1}{C^9} \qquad \frac{1}{C^{11}}$$

Fig. 5.4 We choose x_0^2 in one of the subintervals $\left[\frac{1}{C^{2k-1}} , \frac{1}{C^{2k+1}} \right]$ for some $k \in \mathbb{N}$.

Using similar techniques with subintervals and decomposing the transient terms into three groups of geometric subsequences, we can show that every solution is periodic or eventually periodic with:

- Period-4 if $C > 1$.

- Period-3 if $C = 1$.

Max-Type Difference Equations can be applied in statistics, computer science, electrical engineering, biological sciences, and economics. In the next section, we will advance our studies on periodic and eventually periodic solutions of a Nonautonomous Max-Type Δ.E.

5.5 Eventually Periodic Solutions of Nonautonomous Max-Type Difference Equations

This section's aims are to examine the existence and patterns of periodic and eventually periodic solutions of the following Nonautonomous Max-Type Δ.E.:

$$x_{n+2} = \max\left\{\frac{1}{x_{n+1}}, \frac{C_n}{x_n}\right\}, \ n = 0, 1, \dots, \tag{5.9}$$

where $x_0, x_1 > 0$ and $\{C_n\}_{n=0}^{\infty}$ is a period-2 sequence. In fact, it was shown that every positive solution of Equation (5.9) is eventually periodic with the following periods:

$$\begin{cases} 2 \text{ if } C_0 C_1 < 1, \\ 6 \text{ if } C_0 C_1 = 1, \\ 4 \text{ if } C_0 C_1 > 1. \end{cases}$$

The following examples will portray the patterns of periodic and eventually periodic solutions of Equation (5.9).

Example 23. Suppose that $\{C_n\}_{n=0}^{\infty}$ is a period-2 sequence such that $C_0 C_1 < 1$. Determine the necessary and sufficient conditions for the existence of period-2 solutions of:

$$x_{n+2} = \max\left\{\frac{1}{x_{n+1}}, \frac{C_n}{x_n}\right\}, \ n = 0, 1, \dots.$$

Solution: From Theorem (8), let $x_1 = \frac{1}{x_0}$. Then:

$$x_0,$$

$$x_1 = \frac{1}{x_0},$$

$$x_2 = \max\left\{\frac{1}{[x_1]}, \frac{C_0}{x_0}\right\} = \max\left\{x_0, \frac{C_0}{x_0}\right\} = x_0 \ \left(\text{if } x_0^2 > C_0\right),$$

$$x_3 = \max\left\{\frac{1}{[x_2]}, \frac{C_1}{[x_1]}\right\} = \max\left\{\frac{1}{x_0}, C_1 x_0\right\} = \frac{1}{x_0} = x_1 \,.$$

$$\left(\text{if } x_0^2 < \frac{1}{C_1}\right).$$

Hence we obtain period-2 cycles if and only if:

$$C_0 < x_0^2 < \frac{1}{C_1}\,.$$

Therefore, if either $x_0^2 < C_0$ or $x_0^2 > \frac{1}{C_1}$, then Equation (5.9) will have eventually periodic solutions with transient terms. Similar to Equation (5.8) in the previous section, the pattern of the transient terms will also emerge in three geometrical subsequences.

Example 24. Suppose that $\{C_n\}_{n=0}^{\infty}$ is a period-2 sequence such that $C_0 C_1 < 1$. Determine the necessary and sufficient conditions for the existence of eventually period-2 solutions with three transient terms of:

$$x_{n+2} = \max\left\{\frac{1}{x_{n+1}}, \frac{C_n}{x_n}\right\}\,, \quad n = 0, 1, \ldots\,.$$

Solution: Let $x_1 = \frac{1}{x_0}$. Then:

$$x_0\,,$$

$$x_1 = \frac{1}{x_0}\,,$$

$$x_2 = \max\left\{\frac{1}{[x_1]}, \frac{C_0}{x_0}\right\} = \max\left\{x_0, \frac{C_0}{x_0}\right\} = \frac{C_0}{x_0} \quad (\text{as } x_0^2 < C_0)\,,$$

$$x_3 = \max\left\{\frac{1}{[x_2]}, \frac{C_1}{[x_1]}\right\} = \max\left\{\frac{x_0}{C_0}, C_1 x_0\right\} = \frac{x_0}{C_0} \quad (\text{as } C_0 C_1 < 1)\,,$$

$$x_4 = \max\left\{\frac{1}{[x_3]}, \frac{C_0}{[x_2]}\right\} = \max\left\{\frac{C_0}{x_0}, x_0\right\} = \frac{C_0}{x_0} \quad (\text{as } x_0^2 < C_0)\,,$$

$$x_5 = \max\left\{\frac{1}{[x_4]}, \frac{C_1}{[x_3]}\right\} = \max\left\{\frac{x_0}{C_0}, \frac{C_1 C_0}{x_0}\right\} = \frac{x_0}{C_0} = x_3$$

$$\left(\text{if } (C_0 C_1)C_0 < x_0^2 < C_0\right)\,,$$

$$x_6 = \max\left\{\frac{1}{[x_5]}, \frac{C_0}{[x_4]}\right\} = \max\left\{\frac{C_0}{x_0}, x_0\right\} = \frac{C_0}{x_0} = x_4$$

$$\left(\text{as } x_0^2 < C_0\right).$$

Notice when:

$$[C_0 C_1] C_0 < x_0^2 < C_0 ,$$

we acquire three transient terms in square brackets prior to the period-2 pattern:

$$\left[x_0 , \frac{1}{x_0} , \frac{C_0}{x_0} \right] , \frac{x_0}{C_0} , \frac{C_0}{x_0} , \dots .$$

Example 25. Suppose that $\{C_n\}_{n=0}^{\infty}$ is a period-2 sequence such that $C_0 C_1 < 1$. Determine the necessary and sufficient conditions for the existence of eventually period-2 solutions with six transient terms of:

$$x_{n+2} = \max \left\{ \frac{1}{x_{n+1}} , \frac{C_n}{x_n} \right\} , \quad n = 0, 1, \dots .$$

Solution: Let $x_1 = \frac{1}{x_0}$, then we obtain:

$$x_0 ,$$

$$x_1 = \frac{1}{x_0} ,$$

$$x_2 = \max \left\{ \frac{1}{[x_1]} , \frac{C_0}{x_0} \right\} = \max \left\{ x_0, \frac{C_0}{x_0} \right\} = \frac{C_0}{x_0} \quad \left(\text{as } x_0^2 < C_0 \right) ,$$

$$x_3 = \max \left\{ \frac{1}{[x_2]} , \frac{C_1}{[x_1]} \right\} = \max \left\{ \frac{x_0}{C_0} , C_1 x_0 \right\} = \frac{x_0}{C_0} \quad \left(\text{as } C_0 C_1 < 1 \right) ,$$

$$x_4 = \max \left\{ \frac{1}{[x_3]} , \frac{C_0}{[x_2]} \right\} = \max \left\{ \frac{C_0}{x_0} , x_0 \right\} = \frac{C_0}{x_0} \quad \left(\text{as } x_0^2 < C_0 \right) ,$$

$$x_5 = \max \left\{ \frac{1}{[x_4]} , \frac{C_1}{[x_3]} \right\} = \max \left\{ \frac{x_0}{C_0} , \frac{C_0 C_1}{x_0} \right\} = \frac{C_0 C_1}{x_0}$$

$$\left(\text{as } x_0^2 < [C_0 C_1] C_0 \right) ,$$

$$x_6 = \max \left\{ \frac{1}{[x_5]} , \frac{C_0}{[x_4]} \right\} = \max \left\{ \frac{x_0}{C_0 C_1} , x_0 \right\} = \frac{x_0}{C_0 C_1}$$

$$\left(\text{as } C_0 C_1 < 1 \right) ,$$

$$x_7 = \max \left\{ \frac{1}{[x_6]} , \frac{C_1}{[x_5]} \right\} = \max \left\{ \frac{C_0 C_1}{x_0} , \frac{x_0}{C_0} \right\} = \frac{C_0 C_1}{x_0} = x_5$$

$$\left(\text{as } x_0^2 < [C_0 C_1] C_0 \right) ,$$

$$x_8 = \max\left\{\frac{1}{[x_7]}, \frac{C_0}{[x_6]}\right\} = \max\left\{\frac{x_0}{C_0C_1}, \frac{C_0^2C_1}{x_0}\right\} = \frac{x_0}{C_0C_1} = x_6\,,$$

$$\left(\text{as } [C_0C_1]^2C_0 < x_0^2 < [C_0C_1]C_0\right).$$

Now observe when:

$$[C_0C_1]^2C_0 < x_0^2 < [C_0C_1]C_0\,,$$

we obtain six transient terms in square brackets prior to the period-2 pattern:

$$\left[x_0\,, \frac{1}{x_0}\,, \frac{C_0}{x_0}\,, \frac{x_0}{C_0}\,, \frac{C_0}{x_0}\,, \frac{C_1C_0}{x_0}\right]\,, \frac{x_0}{C_0C_1}\,, \frac{C_0C_1}{x_0}\,, \ldots\,.$$

It is peculiar to observe that the six transient terms can be decomposed into three groups of geometric subsequences.

From Example 24 and Example 25, we can deduce that there exists $k \in \mathbb{N}$ such that:

$$[C_0C_1]^{k+1}C_0 < x_0^2 < [C_0C_1]^kC_0\,,$$

and extends to the following theorem.

Theorem 10. *Suppose that $x_0, x_1 > 0$ and $\{C_n\}_{n=0}^{\infty}$ is a period-2 sequence. Then every solution of Equation (5.9) is eventually periodic with period-2 with 3k transient terms if:*

(1) $C_0C_1 < 1$

(2) There exists $k \in \mathbb{N}$ such that:

$$[C_0C_1]^{k+1}C_0 < x_0^2 < [C_0C_1]^kC_0\,.$$

In addition, using similar techniques with subintervals and decomposing the transient terms into three groups of geometric subsequences, we can show that every solution of Equation (5.9) is periodic or eventually periodic with:

- Period-4 if $C_0C_1 > 1$.

- Period-6 if $C_0C_1 = 1$.

This will be left as an exercise at the end of the chapter. More studies on Max-Type Difference Equations will be addressed in Chapter 6. The following questions will be conveyed about the periodicity nature of solutions:

- When $\{C_n\}_{n=0}^{\infty}$ is a period-k sequence, $(k \geq 3)$.

- Delayed Max-Type Difference Equations.

- Max-Type Difference Equations with three or more components in the form:

$$x_{n+m} = \max\left\{\frac{C_{n+m-1}}{x_{n+m-1}}\,, \ldots\,, \frac{C_1}{x_{n+1}}\,, \frac{C_0}{x_n}\right\}\,, n = 0, 1, \ldots\,,$$

Further properties on the boundedness and periodic nature of solutions of Max-Type Difference Equations have been studied in [26, 29], and [30].

5.6 Exercises

In problems 1–11, suppose that $A, B, C, D > 0$ and $x_0, x_1 \geq 0$. Determine the necessary and sufficient conditions for the existence of periodic solutions and their pattern:

1. $x_{n+2} = \frac{A + Bx_n}{C + Dx_{n+1}}$, $n = 0, 1, \ldots$.

2. $x_{n+2} = \frac{Bx_n}{Cx_{n+1} + Dx_n}$, $n = 0, 1, \ldots$.

3. $x_{n+2} = \frac{Ax_{n+1} + Bx_n}{1 + Cx_{n+1}}$, $n = 0, 1, \ldots$.

4. $x_{n+2} = \frac{Ax_{n+1} + Bx_n}{1 + Cx_n}$, $n = 0, 1, \ldots$.

5. $x_{n+2} = \frac{Ax_{n+1} + Bx_n}{1 + Cx_{n+1} + Dx_n}$, $n = 0, 1, \ldots$.

6. $x_{n+2} = \frac{1 + Ax_{n+1} + Bx_n}{Cx_{n+1} + Dx_n}$, $n = 0, 1, \ldots$.

7. $x_{n+2} = \frac{1}{x_n x_{n+1}}$, $n = 0, 1, \ldots$.

8. $x_{n+2} = \frac{x_n}{x_{n+1}}$, $n = 0, 1, \ldots$.

9. $x_{n+2} = \frac{(-1)^n}{x_n x_{n+1}}$, $n = 0, 1, \ldots$.

10. $x_{n+2} = \frac{(-1)^n x_n}{x_{n+1}}$, $n = 0, 1, \ldots$.

11. $x_{n+2} = \frac{(-1)^n x_{n+1}}{x_n}$, $n = 0, 1, \ldots$.

Consider the **Nonautonomous Rational** Δ.E.:

$$x_{n+2} = \frac{A_n x_n}{1 + x_{n+1}} \quad , \quad n = 0, 1, \ldots .$$

where $\{A_n\}_{n=0}^{\infty}$ is a periodic sequence. In problems 12–23:

12. Determine the necessary and sufficient conditions for the existence and patterns of period-10 solutions when $\{A_n\}_{n=0}^{\infty}$ is a period-5 sequence.

13. Using Exercise 12, determine the necessary and sufficient conditions for the existence and patterns of period-$(2k+1)$ solutions when $\{A_n\}_{n=0}^{\infty}$ is a period-$(2k+1)$ sequence ($k \in \mathbb{N}$).

14. Determine the necessary and sufficient conditions for the existence and patterns of period-2 solutions when $\{A_n\}_{n=0}^{\infty}$ is a period-24 sequence.

15. Determine the necessary and sufficient conditions for the existence and patterns of period-4 solutions when $\{A_n\}_{n=0}^{\infty}$ is a period-24 sequence.

16. Determine the necessary and sufficient conditions for the existence and patterns of period-6 solutions when $\{A_n\}_{n=0}^{\infty}$ is a period-24 sequence.

17. Determine the necessary and sufficient conditions for the existence and patterns of period-12 solutions when $\{A_n\}_{n=0}^{\infty}$ is a period-24 sequence.

18. Determine the necessary and sufficient conditions for the existence and patterns of period-24 solutions when $\{A_n\}_{n=0}^{\infty}$ is a period-24 sequence.

19. Using Exercises 14–18, determine the necessary and sufficient conditions for the existence and pattern of period-2 solutions when $\{A_n\}_{n=0}^{\infty}$ is a period-8k sequence ($k \in \mathbb{N}$).

20. Using Exercises 14–18, determine the necessary and sufficient conditions for the existence and pattern of period-4 solutions when $\{A_n\}_{n=0}^{\infty}$ is a period-8k sequence ($k \in \mathbb{N}$).

21. Using Exercises 14–18, determine the necessary and sufficient conditions for the existence and pattern of period-8 solutions when $\{A_n\}_{n=0}^{\infty}$ is a period-8k sequence ($k \in \mathbb{N}$).

22. Using Exercises 14–18, determine the necessary and sufficient conditions for the existence and pattern of period-2k solutions when $\{A_n\}_{n=0}^{\infty}$ is a period-8k sequence ($k \in \mathbb{N}$).

23. Using Exercises 14–18, determine the necessary and sufficient conditions for the existence and pattern of period-4k solutions when $\{A_n\}_{n=0}^{\infty}$ is a period-8k sequence ($k \in \mathbb{N}$).

Consider the **Nonautonomous Rational** Δ.**E.**:

$$x_{n+2} = \frac{A_n x_n}{1 + x_{n+1} + x_n} \quad , \quad n = 0, 1, \dots .$$

where $\{A_n\}_{n=0}^{\infty}$ is a periodic sequence. In problems 24–31:

24. Determine the necessary and sufficient conditions for the existence and pattern of a period-2 solution when $\{A_n\}_{n=0}^{\infty}$ is a period-12 sequence.

25. Determine the necessary and sufficient conditions for the existence and pattern of a period-4 solution when $\{A_n\}_{n=0}^{\infty}$ is a period-12 sequence.

26. Determine the necessary and sufficient conditions for the existence and pattern of a period-6 solution when $\{A_n\}_{n=0}^{\infty}$ is a period-12 sequence.

27. Determine the necessary and sufficient conditions for the existence and pattern of a period-12 solution when $\{A_n\}_{n=0}^{\infty}$ is a period-12 sequence.

28. Determine the necessary and sufficient conditions for the existence and pattern of a period-2 solution when $\{A_n\}_{n=0}^{\infty}$ is a period-4k sequence (for $k \geq 2$).

29. Determine the necessary and sufficient conditions for the existence and pattern of a period-2k solution when $\{A_n\}_{n=0}^{\infty}$ is a period-4k sequence (for $k \geq 2$).

30. Determine the necessary and sufficient conditions for the existence and pattern of a period-4k solution when $\{A_n\}_{n=0}^{\infty}$ is a period-4k sequence (for $k \geq 2$).

31. Determine the necessary and sufficient conditions for the existence and pattern of a period-$2(2k+1)$ when $\{A_n\}_{n=0}^{\infty}$ is a period-$(2k+1)$ sequence (for $k \geq 1$).

Consider the **Nonautonomous Riccati Δ.E.:**

$$x_{n+2} = \frac{A_n}{x_n} \quad , \quad n = 0, 1, \dots .$$

where $\{A_n\}_{n=0}^{\infty}$ is a periodic sequence. In problems 32–40:

32. Determine the necessary and sufficient conditions for the existence and pattern of period-8 solutions when $\{A_n\}_{n=0}^{\infty}$ is a period-8 sequence.

33. Determine the necessary and sufficient conditions for the existence and pattern of period-12 solutions when $\{A_n\}_{n=0}^{\infty}$ is a period-12 sequence.

33. Using Exercises 32 and 33, determine the necessary and sufficient conditions for the existence and pattern period-4k solutions when $\{A_n\}_{n=0}^{\infty}$ is a period-4k sequence, (for $k \geq 2$).

35. Determine the necessary and sufficient conditions for the existence and pattern of period-10 solution when $\{A_n\}_{n=0}^{\infty}$ is a period-10 sequence.

36. Determine the necessary and sufficient conditions for the existence and pattern of period-14 solution when $\{A_n\}_{n=0}^{\infty}$ is a period-14 sequence.

37. Using Exercises 35 and 36, determine the necessary and sufficient conditions for the existence and pattern of period-$(4k+2)$ solution when $\{A_n\}_{n=0}^{\infty}$ is a period-$(4k+2)$ sequence, (for $k \geq 2$).

38. Determine the necessary and sufficient conditions for the existence and pattern of period-5 solution when $\{A_n\}_{n=0}^{\infty}$ is a period-5 sequence. **Hint:** set $x_{10} = x_0$ and $x_{11} = x_1$.

39. Determine the necessary and sufficient conditions for the existence and pattern of period-7 solution when $\{A_n\}_{n=0}^{\infty}$ is a period-7 sequence. **Hint:** set $x_{14} = x_0$ and $x_{15} = x_1$.

40. Using Exercises 38 and 39, determine the necessary and sufficient conditions for the existence of the period-$(2k+1)$ solution when $\{A_n\}_{n=0}^{\infty}$ is a period-$(2k+1)$ sequence, $(k \in \mathbb{N})$. **Hint:** Break up into two cases when $\{A_n\}_{n=0}^{\infty}$ is a period-$(4k-1)$ sequence and when $\{A_n\}_{n=0}^{\infty}$ is a period-$(4k+1)$ sequence, $(k \in \mathbb{N})$.

Consider the **Max-Type Δ.E.**:

$$x_{n+2} = \max\left\{\frac{1}{x_{n+1}}, \frac{C}{x_n}\right\}, \quad n = 0, 1, \dots,$$

where $C > 0$. In problems 41–52:

41. Suppose that $C < 1$, and $C^7 < x_0^2 < C^5$. Determine the pattern of the transient terms and the period-2 solutions.

42. Suppose that $C < 1$, and $C^9 < x_0^2 < C^7$. Determine the pattern of the transient terms and the period-2 solutions.

43. Using Exercises 41 and 42, suppose that $C < 1$, and $C^{2k+1} < x_0^2 < C^{2k-1}$ $(k \in \mathbb{N})$. Determine the pattern of the transient terms and the period-2 solutions.

44. Suppose that $C < 1$, and $C < x_0^2 < \frac{1}{C}$. Determine the pattern of the transient terms and the period-2 solutions.

45. Suppose that $C < 1$, and $\frac{1}{C} < x_0^2 < \frac{1}{C^3}$. Determine the pattern of the transient terms and the period-2 solutions.

46. Suppose that $C < 1$, and $\frac{1}{C^3} < x_0^2 < \frac{1}{C^5}$. Determine the pattern of the transient terms and the period-2 solutions.

47. Using Exercises 44, 45, and 46, suppose that $C < 1$, and $\frac{1}{C^{2k-1}} < x_0^2 < \frac{1}{C^{2k+1}}$ (for $k \geq 1$). Determine the pattern of the transient terms and the period-2 solutions.

48. Suppose that $C > 1$. Determine the necessary and sufficient conditions for the existence of the period-4 solutions.

49. Suppose that $C > 1$. Determine the necessary and sufficient conditions for the existence of the period-4 solutions with three transient terms.

50. Suppose that $C > 1$. Determine the necessary and sufficient conditions for the existence of the period-4 solutions with six transient terms.

51. Suppose that $C > 1$. Determine the necessary and sufficient conditions for the existence of the period-4 solutions with nine transient terms.

52. Using Exercises 49, 50, and 51, suppose that $C > 1$. Determine the pattern of the 3k transient terms $(k \in \mathbb{N})$ and the period-4 solutions.

Consider the **Nonautonomous Max-Type Δ.E.:**

$$x_{n+2} = \max\left\{\frac{1}{x_{n+1}}, \frac{C_n}{x_n}\right\}, \; n = 0, 1, \ldots,$$

where $\{C_n\}_{n=0}^{\infty}$ is a period-2 sequence. In problems 53–61:

53. Suppose that $C_0 C_1 < 1$, and $[C_0 C_1]^3 C_0 < x_0^2 < [C_0 C_1]^2 C_0$. Determine the pattern of the transient terms and the period-2 solutions.

54. Suppose that $C_0 C_1 < 1$, and $[C_0 C_1]^4 C_0 < x_0^2 < [C_0 C_1]^3 C_0$. Determine the pattern of the transient terms and the period-2 solutions.

55. Suppose that $C_0 C_1 < 1$, and $[C_0 C_1]^5 C_0 < x_0^2 < [C_0 C_1]^4 C_0$. Determine the pattern of the transient terms and the period-2 solutions.

56. Using Exercises 53, 54, and 55 suppose that $C_0 C_1 < 1$, and $[C_0 C_1]^{k+1} C_0 < x_0^2 < [C_0 C_1]^k C_0$ (for $k \geq 1$). Determine the pattern of the transient terms and the period-2 solutions.

57. Suppose that $C_0 C_1 > 1$. Determine the necessary and sufficient conditions for the existence of the period-4 solutions.

58. Suppose that $C_0 C_1 > 1$. Determine the necessary and sufficient conditions for the existence of the period-4 solutions with three transient terms.

59. Suppose that $C_0 C_1 > 1$. Determine the necessary and sufficient conditions for the existence of the period-4 solutions with six transient terms.

60. Suppose that $C_0 C_1 > 1$. Determine the necessary and sufficient conditions for the existence of the period-4 solutions with nine transient terms.

61. Using Exercises 58, 59, and 60, suppose that $C_0 C_1 > 1$. Determine the pattern of the 3k transient terms ($k \in \mathbb{N}$) and the period-4 solutions.

Chapter 6
Advanced Periodic Characteristics and New Research Questions

6.1 Advanced Periodic Characteristics

In the previous chapters we explored varieties of periodic traits of linear difference equations, piece-wise difference equations, rational difference equations, and Max-Type difference equations. First of all, we observed how the periodicity nature varied from the existence of unique periodic cycles to every solution being periodic. Second, we discovered the necessary and sufficient conditions for the existence of periodic cycles. Furthermore, we perceived the disparity between patterns of the even ordered periodic cycles and odd ordered periodic cycles with uniqueness of solutions and alternating patterns. Our next aim is to broaden our developed knowledge to ascertain the periodicity essence of third and higher order difference equations, systems of difference equations, when the periodic sequence $\{A_n\}_{n=0}^{\infty}$ is periodic with period-k, ($k \geq 3$) and study of new systems of piecewise difference equations and neural networking models. How will these criteria for the existence of periodic cycles be similar or different compared to our current mastery in previous chapters? What new periodic traits can we expect that will be quite different? How will the delay $m \geq 3$ affect the new periodic traits? Will even and odd values of delay $m \geq 3$ influence the new periodic traits? We will pose conjectures that will lead to new theorems and results based on our previous studies and will pose new research questions.

6.2 Third and Higher Order Linear Difference Equations

From Chapter 4, recall a homogeneous linear Δ.E. of order $m \geq 3$ in the form:

$$x_{n+m} + \sum_{i=1}^{m} a_i x_{n+i-1} = 0 \quad , \quad n = 0, 1, \dots . \tag{6.1}$$

© Springer Nature Switzerland AG 2018
M. A. Radin, *Periodic Character and Patterns of Recursive Sequences*,
https://doi.org/10.1007/978-3-030-01780-4_6

It is of paramount interest to determine the periodic character and patterns and the necessary and sufficient conditions for the existence of periodic cycles of Equation (6.1). The periodic traits of Equation (6.1) have been studied by [15, 16, 17] and [18]. We will begin our journey by suggesting the following **Open Problems** similar to the ones in Chapter 2 and in Chapter 4.

Open Problem 1. *Suppose that $m \geq 3$ and $\{a_n\}_{n=0}^{\infty}$ is a period-k sequence, ($k \geq 2$). Determine the necessary and sufficient conditions for the existence of periodic solutions of:*

$$x_{n+m} = x_n + a_n \quad , \quad n = 0, 1, \dots .$$

Open Problem 2. *Suppose that $m \geq 3$ and $\{a_n\}_{n=0}^{\infty}$ is a period-k sequence, ($k \geq 2$). Determine the necessary and sufficient conditions for the existence of periodic solutions of:*

$$x_{n+m} = -x_n + a_n \quad , \quad n = 0, 1, \dots .$$

Open Problem 3. *Suppose that $m \geq 3$ and $\{a_n\}_{n=0}^{\infty}$ is a period-k sequence, ($k \geq 2$). Determine the necessary and sufficient conditions for the existence of periodic solutions of:*

$$x_{n+m} = a_n x_n \quad , \quad n = 0, 1, \dots .$$

Open Problem 4. *Suppose that $m \geq 3$ and $\{a_n\}_{n=0}^{\infty}$ is a period-k sequence, ($k \geq 2$). Determine the necessary and sufficient conditions for the existence of periodic solutions of:*

$$x_{n+m} = -a_n x_n \quad , \quad n = 0, 1, \dots .$$

Open Problem 5. *Suppose that $m \geq 3$ and $\{a_n\}_{n=0}^{\infty}$ and $\{b_n\}_{n=0}^{\infty}$ are period-k sequences, ($k \geq 2$). Determine the necessary and sufficient conditions for the existence of periodic solutions of:*

$$x_{n+m} = a_n x_n + b_n \quad , \quad n = 0, 1, \dots .$$

Open Problem 6. *Suppose that $m \geq 3$ and $\{a_n\}_{n=0}^{\infty}$ and $\{b_n\}_{n=0}^{\infty}$ are period-k sequences, ($k \geq 2$). Determine the necessary and sufficient conditions for the existence of periodic solutions of:*

$$x_{n+m} = -a_n x_n + b_n \quad , \quad n = 0, 1, \dots .$$

Hint: To establish theorems from Open Problems (1)–(6), perform several examples with different values of m and k. Decompose into four cases when k and m are both even, are both odd, and when k is even and m is odd and vice versa. For instance, the following theorem is an example of Open Problem (3).

Suppose $\{a_n\}_{n=0}^{\infty}$ is a period-k sequence, ($k \geq 2$) and consider the nonautonomous third order linear Δ.E.:

$$x_{n+3} = a_n x_n \quad , \quad n = 0, 1, \dots , \tag{6.2}$$

where $x_0 \neq 0$. The following theorem retorts the questions addressed in Open Problem (3) and outlines the periodicity character of solutions of Equation (6.2).

Theorem 7. *Suppose that* $\{a_n\}_{n=0}^{\infty}$ *is a period-k sequence,* $(k \geq 2)$ *and* $x_0 \neq 0$. *Then the following statements are true:*

(i) *If* $\{a_n\}_{n=0}^{\infty}$ *is a period-3 sequence, then Equation (6.2) has no period-3 cycles.*

(ii) *If* $\{a_n\}_{n=0}^{\infty}$ *is a period-3 sequence, then every solution of Equation (6.2) is periodic with period-6 if and only if* $a_0 = \pm 1$, $a_1 = \pm 1$ *and* $a_2 = \pm 1$ *and are not all equal to each other.*

(iii) *If* $\{a_n\}_{n=0}^{\infty}$ *is a period-3k sequence* $(k \geq 2)$, *then every solution of Equation (6.2) is periodic with period-3k if and only if for all* $j = 0, 1, 2$:

$$\prod_{i=0}^{k-1} a_{3i+j} = 1.$$

(iv) *If* $\{a_n\}_{n=0}^{\infty}$ *is a period-$(3k+1)$ sequence* $(k \in \mathbb{N})$, *then every solution of Equation (6.2) is periodic with period-$3(3k+1)$ if and only if:*

$$\prod_{i=0}^{3k} a_i = 1.$$

(v) *If* $\{a_n\}_{n=0}^{\infty}$ *is a period-$(3k+2)$ sequence* $(k \geq 0)$, *then every solution of Equation (6.2) is periodic with period-$3(3k+2)$ if and only if:*

$$\prod_{i=0}^{3k+1} a_i = 1.$$

Theorem (7) is proved by induction.

Now suppose $\{a_n\}_{n=0}^{\infty}$ is a period-k sequence, $(k \geq 2)$ and consider the nonautonomous linear Δ.E. of order m $(m \geq 3)$ in the form:

$$x_{n+m} = a_n x_n \quad, \quad n = 0, 1, \ldots, \tag{6.3}$$

where $x_0 \neq 0$. Then Theorem (7) extends to the following theorem.

Theorem 8. *Suppose that* $\{a_n\}_{n=0}^{\infty}$ *is a period-k sequence,* $(k \geq 2)$ *and* $x_0 \neq 0$. *Then the following statements are true:*

(i) *If* $\{a_n\}_{n=0}^{\infty}$ *is a period-m sequence* $(m \geq 3)$, *then Equation (6.3) has no period-m cycles.*

(ii) *If* $\{a_n\}_{n=0}^{\infty}$ *is a period-m sequence, then every solution of Equation (6.3) is periodic with period-2m if and only if* $a_i = \pm 1$ *for all* $i = 0, 1, \ldots, m-1$ *and are not all equal to each other.*

(iii) If $\{a_n\}_{n=0}^{\infty}$ is a period-mk sequence ($k \geq 2$), then every solution of Equation (6.3) is periodic with period-mk if and only if for all $j = 0, 1, \ldots, m-1$:

$$\prod_{i=0}^{k-1} a_{mi+j} = 1 .$$

(iv) If $\{a_n\}_{n=0}^{\infty}$ is a period-$(mk+j)$ ($j = 1, \ldots, m-1$), then every solution of Equation (6.3) is periodic with period-$m(mk+j)$ if and only if:

$$\prod_{i=0}^{mk+j-1} a_i = 1 .$$

We will shift our focus on Open Problem (1). Suppose $\{a_n\}_{n=0}^{\infty}$ is a period-k sequence, ($k \geq 2$) and consider the nonautonomous third order linear Δ.E.:

$$x_{n+3} = x_n + a_n \ , \quad n = 0, 1, \ldots , \tag{6.4}$$

The following theorem addresses the questions in Open Problem (1) and the periodicity character of Equation (6.4).

Theorem 9. *Suppose that $\{a_n\}_{n=0}^{\infty}$ is a period-k sequence, ($k \geq 2$) and $x_0 \in \mathfrak{R}$. Then the following statements are true:*

(i)   *If $\{a_n\}_{n=0}^{\infty}$ is a period-3 sequence, then Equation (6.4) has no period-3 cycles.*

(ii) *If $\{a_n\}_{n=0}^{\infty}$ is a period-3k sequence ($k \geq 2$), then every solution of Equation (6.4) is periodic with period-3k if and only if for all $j = 0, 1, 2$:*

$$\sum_{i=0}^{k-1} a_{3i+j} = 0 .$$

(iii) *If $\{a_n\}_{n=0}^{\infty}$ is a period-$(3k+1)$ sequence ($k \in \mathbb{N}$), then every solution of Equation (6.4) is periodic with period-$3(3k+1)$ if and only if:*

$$\sum_{i=0}^{3k} a_i = 0 .$$

(v)   *If $\{a_n\}_{n=0}^{\infty}$ is a period-$(3k+2)$ sequence ($k \geq 0$), then every solution of Equation (6.4) is periodic with period-$3(3k+2)$ if and only if:*

$$\sum_{i=0}^{3k+1} a_i = 0 .$$

6.3 Systems of Linear Difference Equations

Our next goal is to establish the patterns of periodic cycles of particular systems of linear difference equations. We will emerge with several examples of various systems of linear difference equations and study their periodic traits.

Example 1. Determine the necessary and sufficient conditions for the existence of periodic solutions of:

$$
\begin{cases}
x_{n+2} = (-1)^n y_n \\
\\
y_{n+2} = x_n
\end{cases}
\qquad n = 0, 1, 2, \ldots .
\tag{6.5}
$$

Solution: By iteration we get:

$$x_0 \qquad\qquad y_0$$

$$x_1 \qquad\qquad y_1$$

$$x_2 = y_0 \qquad\qquad y_2 = x_0$$

$$x_3 = -y_1 \qquad\qquad y_3 = x_1$$

$$x_4 = y_2 = x_0 \qquad\qquad y_4 = y_2 = x_0 .$$

Hence every solution of System (6.5) is periodic with period-4.

Example 2. Suppose that $\{a_n\}_{n=0}^{\infty}$ and $\{b_n\}_{n=0}^{\infty}$ are period-2 sequences. Determine the necessary and sufficient conditions for the existence of periodic solutions of:

$$
\begin{cases}
x_{n+1} = a_n y_n \\
\\
y_{n+1} = b_n x_n
\end{cases}
\qquad n = 0, 1, 2, \ldots .
\tag{6.6}
$$

Solution: Observe:

$$x_0 \qquad\qquad\qquad y_0$$

$$x_1 = a_0 y_0 \qquad\qquad\qquad y_1 = b_0 x_0$$

$$x_2 = a_1 [y_1] = a_1 b_0 x_0 = x_0 \qquad y_2 = b_1 [x_1] = b_1 a_0 y_0 = y_0$$

$$(\text{if } a_1 b_0 = 1). \qquad\qquad (\text{if } b_1 a_0 = 1).$$

Thus every solution of System (6.6) is periodic with period-2 if and only if $a_1 b_0 = 1$ and $b_1 a_0 = 1$.

Example 3. Suppose that $\{a_n\}_{n=0}^{\infty}$ and $\{b_n\}_{n=0}^{\infty}$ are period-3 sequences. Determine the necessary and sufficient conditions for the existence of periodic solutions of:

$$\begin{cases} x_{n+1} = a_n y_n \\ \\ y_{n+1} = b_n x_n \end{cases} \quad n = 0, 1, 2, \ldots . \tag{6.7}$$

Solution: Notice:

x_0

$x_1 = a_0 y_0$

$x_2 = a_1 [y_1] = a_1 b_0 x_0$

$x_3 = a_2 [y_2] = a_2 b_1 a_0 y_0 = y_0$

(if $a_2 b_1 a_0 = 1$)

$x_4 = a_0 [y_3] = a_0 y_3$

$x_5 = a_1 [y_4] = a_1 b_0 x_3$

$x_6 = a_2 [y_5] = a_2 b_1 a_0 y_3 = y_3 = x_0$

y_0

$y_1 = b_0 x_0$

$y_2 = b_1 [x_1] = b_1 a_0 y_0$

$y_3 = b_2 [x_2] = b_2 a_1 b_0 x_0 = x_0$

(if $b_2 a_1 b_0 = 1$)

$y_4 = b_0 [x_3] = b_0 x_3$

$y_5 = b_1 [x_4] = b_1 a_0 y_3$

$y_6 = b_2 [x_5] = b_2 a_1 b_0 x_3 = x_3 = y_0$.

Every solution of System (6.7) is periodic with period-6 if and only if $a_2 b_1 a_0 = 1$ and $b_2 a_1 b_0 = 1$.

In Examples (2) and (3) we can see contrasts when $\{a_n\}_{n=0}^{\infty}$ and $\{b_n\}_{n=0}^{\infty}$ are both periodic sequences with an even ordered period and with an odd ordered period. The following two theorems generalize these observations.

Theorem 10. *Suppose that $\{a_n\}_{n=0}^{\infty}$ and $\{b_n\}_{n=0}^{\infty}$ are periodic sequences with period-2k, ($k \in \mathbb{N}$). Then every solution of:*

$$\begin{cases} x_{n+1} = a_n y_n \\ \\ \\ y_{n+1} = b_n x_n \end{cases} \quad n = 0, 1, 2, \ldots ,$$

is periodic with period-2k if and only if

$$\prod_{i=1}^{k} a_{2i-1}b_{2i-2} = 1 \quad \text{and} \quad \prod_{i=1}^{k} b_{2i-1}a_{2i-2} = 1 \, .$$

To prove Theorem 10, test several examples when $k = 2, 4, \ldots$, and then inductively generalize.

Theorem 11. *Suppose that* $\{a_n\}_{n=0}^{\infty}$ *and* $\{b_n\}_{n=0}^{\infty}$ *are periodic sequences with* period$-(2k+1)$, $(k \in \mathbb{N})$. *Then every solution of:*

$$\begin{cases} x_{n+1} = a_n y_n \\ \\ \\ y_{n+1} = b_n x_n \end{cases} \quad n = 0, 1, 2, \ldots,$$

is periodic with period $-2(2k+1)$ *if and only if*

$$\prod_{i=1}^{k} a_{2i-2}b_{2i-1} = 1 \quad \text{and} \quad \prod_{i=1}^{k} b_{2i-2}a_{2i-1} = 1 \, .$$

To prove Theorem 11, implement various examples when $k = 3, 5 \ldots$, and then inductively proceed. We will advance with more examples systems of difference equations and their periodicity characters.

Example 4. Suppose that $\{a_n\}_{n=0}^{\infty}$ and $\{b_n\}_{n=0}^{\infty}$ are period-2 sequences. Determine the necessary and sufficient conditions for the existence of period-2 solutions of:

$$\begin{cases} x_{n+1} = (-1)^n \, y_n + a_n \\ \\ \\ y_{n+1} = (-1)^n \, x_n + b_n \end{cases} \quad n = 0, 1, 2, \ldots \, . \quad (6.8)$$

Solution: Observe:

$$x_0 \qquad\qquad\qquad y_0$$

$$x_1 = y_0 + a_0 \qquad\qquad y_1 = x_0 + b_0$$

$$x_2 = -[y_1] + a_1 \qquad\qquad y_2 = -[x_1] + b_1$$

$$= -x_0 - b_0 + a_1 = x_0 \qquad = -y_0 - a_0 + b_1 = y_0 \, .$$

Hence $x_0 = \frac{a_1 - b_0}{2}$ and $y_0 = \frac{b_1 - a_0}{2}$ and the following period-2 pattern:

$$x_0 = \frac{a_1 - b_0}{2} \qquad y_0 = \frac{b_1 - a_0}{2}$$

$$x_1 = \frac{b_1 + a_0}{2} \qquad y_1 = \frac{a_1 + b_0}{2} \ .$$

Example 5. Suppose that $\{a_n\}_{n=0}^{\infty}$ and $\{b_n\}_{n=0}^{\infty}$ are period-4 sequences. Determine the necessary and sufficient conditions for the existence of period-4 solutions of:

$$\begin{cases} x_{n+1} = (-1)^n y_n + a_n \\ \\ \\ y_{n+1} = (-1)^n x_n + b_n \end{cases} \qquad n = 0, 1, 2, \ldots . \qquad (6.9)$$

Solution: We obtain:

x_0	y_0
$x_1 = y_0 + a_0$	$y_1 = x_0 + b_0$
$x_2 = -[y_1] + a_1$	$y_2 = -[x_1] + b_1$
$\quad = -x_0 - b_0 + a_1$	$\quad = -y_0 - a_0 + b_1$
$x_3 = [y_2] + a_2$	$y_3 = [x_2] + b_2$
$\quad = -y_0 - a_0 + b_1 + a_2$	$\quad = -x_0 - b_0 + a_1 + b_2$
$x_4 = -[y_3] + a_3$	$y_4 = -[x_3] + b_3$
$\quad = x_0 + b_0 - a_1 - b_2 + a_3$	$\quad = y_0 + a_0 - b_1 - a_2 + b_3$
$\quad = x_0$	$\quad = y_0 \ .$

Thus every solution of System (6.9) is periodic with period-4 if and only if

$$b_0 - b_2 = a_1 - a_3 \quad \text{and} \quad a_0 - a_2 = b_1 - b_3 \ .$$

In Examples 4 and 5 there are variations when $\{a_n\}_{n=0}^{\infty}$ and $\{b_n\}_{n=0}^{\infty}$ are both periodic sequences with period-$(4k - 2)$ and period-$4k$, $(k \in \mathbb{N})$. The following two theorems describe the results.

Theorem 12. *Suppose that* $\{a_n\}_{n=0}^{\infty}$ *and* $\{b_n\}_{n=0}^{\infty}$ *are periodic sequences with period-$(4k-2)$, $(k \in \mathbb{N})$. Then:*

$$
\begin{cases}
x_{n+1} = (-1)^n y_n + a_n \\
\\
\qquad\qquad\qquad\qquad n = 0,1,2,\ \ldots, \\
\\
y_{n+1} = (-1)^n x_n + b_n
\end{cases}
$$

has a unique period-$(4k-2)$ cycle with:

$$
x_0 = \frac{\sum_{i=1}^{2k-1} (-1)^{n+1} a_{2i-1} + \sum_{i=1}^{2k-1} (-1)^n b_{2i-2}}{2}, \text{ and}
$$

$$
y_0 = \frac{\sum_{i=1}^{2k-1} (-1)^{n+1} b_{2i-1} + \sum_{i=1}^{2k-1} (-1)^n a_{2i-2}}{2}.
$$

To prove Theorem 12, seek more examples when $k = 6, 10, \ldots$.

Theorem 13. *Suppose that* $\{a_n\}_{n=0}^{\infty}$ *and* $\{b_n\}_{n=0}^{\infty}$ *are periodic sequences with period-$4k$, $(k \in \mathbb{N})$. Then every solution of:*

$$
\begin{cases}
x_{n+1} = (-1)^n y_n + a_n \\
\\
\qquad\qquad\qquad\qquad n = 0,1,2,\ \ldots, \\
\\
y_{n+1} = (-1)^n x_n + b_n
\end{cases}
$$

is periodic with period-$4k$ if and only if:

$$
\sum_{i=1}^{2k} (-1)^{n+1} b_{2i-2} = \sum_{i=1}^{2k} (-1)^{n+1} a_{2i-1}, \text{ and}
$$

$$
\sum_{i=1}^{2k} (-1)^{n+1} a_{2i-2} = \sum_{i=1}^{2k} (-1)^{n+1} b_{2i-1}.
$$

To prove Theorem (13), endeavor more examples when $k = 8, 12, \ldots$.

6.4 Third and Higher Order Rational Difference Equations

From Chapter 3 and Chapter 5 we investigated the periodic character of solutions of various nonlinear difference equations such as the Riccati Δ.E., piece-wise difference equations, Max-Type Difference Equations, and Rational Difference Equations. We will commence with the investigation of periodic character of third and higher order **Riccati Δ.E.** in the form:

$$x_{n+m} = \frac{A_n}{x_n} \quad , \quad n = 0, 1, \dots , \tag{6.10}$$

where $m \geq 3$ and $\{A_n\}_{n=0}^{\infty}$ is a period-k sequence, $(k \geq 2)$. Equation (6.10) will then guide us to the following **Open Problems** remitting the periodic traits of solutions.

Open Problem 14. *Suppose that m=2l, ($l \in \mathbb{N}$) and $\{A_n\}_{n=0}^{\infty}$ is a period-2k sequence, ($k \in \mathbb{N}$). Determine the necessary and sufficient conditions for the existence of periodic solutions of Equation (6.10).*

Open Problem 15. *Suppose that m=2l, ($l \in \mathbb{N}$) and $\{A_n\}_{n=0}^{\infty}$ is a period-(2k+1) sequence, ($k \in \mathbb{N}$). Determine the necessary and sufficient conditions for the existence of periodic solutions of Equation (6.10).*

Open Problem 16. *Suppose that $m = 2l+1$, ($l \in \mathbb{N}$) and $\{A_n\}_{n=0}^{\infty}$ is a period-2k sequence, ($k \in \mathbb{N}$). Determine the necessary and sufficient conditions for the existence of periodic solutions of Equation (6.10).*

Open Problem 17. *Suppose that $m = 2l + 1$, ($l \in \mathbb{N}$) and $\{A_n\}_{n=0}^{\infty}$ is a period-(2k + 1) sequence, ($k \in \mathbb{N}$). Determine the necessary and sufficient conditions for the existence of periodic solutions of Equation (6.10).*

We will proceed with the analysis of periodicity properties of the nonautonomous rational Δ.E. in the form:

$$x_{n+m} = \frac{A_n x_n}{1 + x_{n+1}} \quad , \quad n = 0, 1, \dots , \tag{6.11}$$

where $m \geq 3$ and $\{A_n\}_{n=0}^{\infty}$ is a period-k sequence, $(k \geq 2)$. Equation (6.11) will then lead to the following **Open Problems** describing the periodic nature of solutions.

Open Problem 18. *Suppose that m=2l, ($l \in \mathbb{N}$) and $\{A_n\}_{n=0}^{\infty}$ is a period-2k sequence, ($k \in \mathbb{N}$). Determine the necessary and sufficient conditions for the existence of periodic solutions of Equation (6.11).*

Open Problem 19. *Suppose that m=2l, ($l \in \mathbb{N}$) and $\{A_n\}_{n=0}^{\infty}$ is a period-(2k+1) sequence, ($k \in \mathbb{N}$). Determine the necessary and sufficient conditions for the existence of periodic solutions of Equation (6.11).*

Open Problem 20. *Suppose that $m = 2l+1$, ($l \in \mathbb{N}$) and $\{A_n\}_{n=0}^{\infty}$ is a period-2k sequence, ($k \in \mathbb{N}$). Determine the necessary and sufficient conditions for the existence of periodic solutions of Equation (6.11).*

Open Problem 21. *Suppose that* $m = 2l + 1$, $(l \in \mathbb{N})$ *and* $\{A_n\}_{n=0}^\infty$ *is a period-*$(2k+1)$ *sequence,* $(k \in \mathbb{N})$.*Determine the necessary and sufficient conditions for the existence of periodic solutions of Equation (6.11).*

We will advance with the study of periodic character of the nonautonomous rational Δ.E. in the form:

$$x_{n+m} = \frac{A_n x_n}{1 + \sum_{i=i}^{m} x_{n+m-i}} \quad , \quad n = 0, 1, \dots ,\qquad (6.12)$$

where $m \geq 3$ and $\{A_n\}_{n=0}^\infty$ is a period-k sequence, $(k \geq 2)$. Equation (6.12) will then direct us to the following **Open Problems** portraying the periodic nature of solutions.

Open Problem 22. *Suppose that* $m=2l$, $(l \in \mathbb{N})$ *and* $\{A_n\}_{n=0}^\infty$ *is a period-2k sequence,* $(k \in \mathbb{N})$. *Determine the necessary and sufficient conditions for the existence of periodic solutions of Equation (6.12).*

Open Problem 23. *Suppose that* $m=2l$, $(l \in \mathbb{N})$ *and* $\{A_n\}_{n=0}^\infty$ *is a period-*$(2k+1)$ *sequence,* $(k \in \mathbb{N})$. *Determine the necessary and sufficient conditions for the existence of periodic solutions of Equation (6.12).*

Open Problem 24. *Suppose that* $m = 2l+1$, $(l \in \mathbb{N})$ *and* $\{A_n\}_{n=0}^\infty$ *is a period-2k sequence,* $(k \in \mathbb{N})$. *Determine the necessary and sufficient conditions for the existence of periodic solutions of Equation (6.12).*

Open Problem 25. *Suppose that* $m = 2l + 1$, $(l \in \mathbb{N})$ *and* $\{A_n\}_{n=0}^\infty$ *is a period-*$(2k+1)$ *sequence,* $(k \in \mathbb{N})$. *Determine the necessary and sufficient conditions for the existence of periodic solutions of Equation (6.12)*

6.5 More About Max-Type Difference Equations

In Chapter 5 we investigated the periodic traits of Max-Type Δ.E. in the form:

$$x_{n+2} = \max\left\{\frac{1}{x_{n+1}}, \frac{C_n}{x_n}\right\}, \, n = 0, 1, \dots ,$$

where $\{C_n\}_{n=0}^\infty$ is a period-2 sequence. Our next aim is to expand the exploration of periodic and eventually periodic solutions when $\{C_n\}_{n=0}^\infty$ is a period-p sequence, $(p \geq 3)$. This will then guide us to the investigation of periodic character of:

$$x_{n+2} = \max\left\{\frac{B_n}{x_{n+1}}, \frac{C_n}{x_n}\right\}, \ n = 0, 1, \dots,$$

where $\{B_n\}_{n=0}^{\infty}$ and $\{C_n\}_{n=0}^{\infty}$ periodic sequences. Furthermore, to study the periodic nature of:

$$x_{n+m} = \max\left\{\frac{B_n}{x_{n+m-l}}, \frac{C_n}{x_{n+m-k}}\right\}, \ n = 0, 1, \dots,$$

where $m \geq 3$, $0 \leq k, l < m$, and $k \neq l$.

6.6 More on Piece-wise Difference Equations

In Chapter 3 we studied the Tent-Map and other piece-wise difference equations; in particular, the Piece-wise Δ.E. in the form:

$$x_{n+1} = \beta_n x_n - g(x_n) \ , \ n = 0, 1, 2, \dots,$$

where $x_0 \in \mathfrak{R}$, $\{\beta_n\}_{n=0}^{\infty}$ is a period-2 sequence and

$$g(x) = \begin{cases} 1 & \text{if } x \geq 0, \\ -1 & \text{if } x < 0. \end{cases}$$

Our next goal is to establish the periodic traits of solutions when $\{\beta_n\}_{n=0}^{\infty}$ is a period-k sequence, $(k \geq 3)$. The following graph (Figure 6.1):

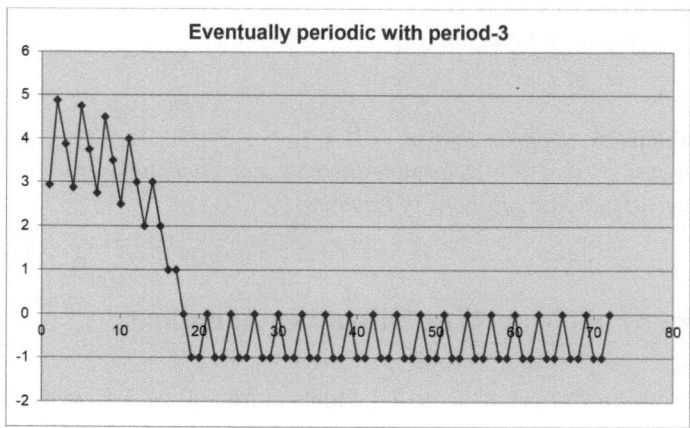

Fig. 6.1 is an example of an eventually periodic cycle with period-3 with 15 transient terms.

We will then constitute the following conjecture.

Conjecture 1. Consider the Δ.E.:

$$x_{n+1} = \beta_n x_n - g(x_n) \ , \ n = 0, 1, 2, \dots, \tag{6.13}$$

where $x_0 \in \mathfrak{R}$, $\{\beta_n\}_{n=0}^{\infty}$ is a period-k sequence, $k \geq 2$ and

$$g(x) = \begin{cases} 1 & \text{if } x \geq 0, \\ -1 & \text{if } x < 0. \end{cases}$$

Then the following statements hold true:

(1) Equation (6.13) has cycles with period kN, for all $N \in \mathbb{N}$.

(2) Equation (6.13) has eventually periodic solutions with period kN, for all $N \in \mathbb{N}$.

Now we will advance with investigating the periodic character of solutions of the following system of piece-wise difference equations:

$$\begin{cases} x_{n+1} = |x_n| - y_n - 1 \\ \\ \\ y_{n+1} = x_n + |y_n| \end{cases} \quad n = 0, 1, 2, \ldots,$$

This system of piece-wise difference equations was investigated in [22] and [31] and showed that there are two period-5 cycles. The above system can be extended to the following system of piece-wise difference equations:

$$\begin{cases} x_{n+1} = |x_n| + ay_n + b \\ y_{n+1} = x_n + c|y_n| + d \end{cases} \quad n = 0, 1, 2, \ldots,$$

where the coefficients a, b, c, and d are either $-1, 0$ or 1. Furthermore, systems of piece-wise difference equations have been used in modeling neural networks. The following system of piece-wise difference equations (**Rulkov Model**) [22] and [31] has been used in neural networking:

$$\left. \begin{array}{l} x_{n+1} = \frac{\alpha}{1 - x_n} + y_n \\ \\ y_{n+1} = y_n - \mu(x_n + 1) + \mu\sigma \end{array} \right\}, \quad n = 0, 1, \ldots \qquad (6.14)$$

when $x_n \leq 0$,

$$\left. \begin{array}{l} x_{n+1} = \alpha + y_n \\ \\ y_{n+1} = y_n - \mu(x_n + 1) + \mu\sigma \end{array} \right\}, \quad n = 0, 1, \ldots \qquad (6.15)$$

when $x_n \leq \alpha + y_n$ and $x_{n-1} \leq 0$,

$$\left.\begin{array}{l} x_{n+1} = -1 \\[2mm] y_{n+1} = y_n - \mu(x_n+1) + \mu\sigma \end{array}\right\}, \quad n = 0, 1, \ldots \qquad (6.16)$$

when $x_n > \alpha + y_n$ or $x_{n-1} > 0$. We will discover quite contrasting periodic behaviors ranging from very steady periodic cycles, periodic cycles with different spikes, eventually periodic behavior, and irregular periodic cycles or chaotic orbits. Similar to piecewise difference equations as the Tent-Map and the Neuron Model in Chapter 3, we have strong sensitivity to the initial conditions and the relationship between the parameters. We will present several graphical examples depicting various periodic characters that describe the **Rulkov Model** [[33, 34]] (Figures 6.2–6.5):

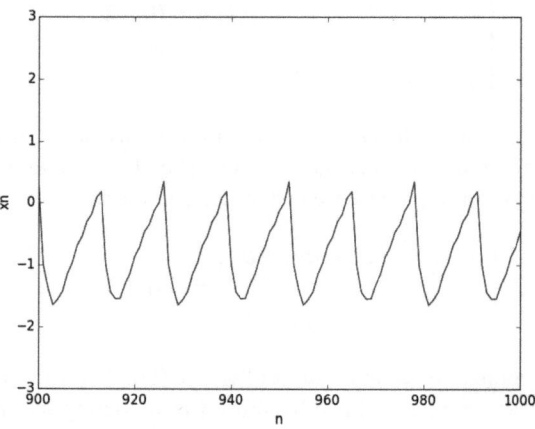

Fig. 6.2 An example of a periodic cycle with a steady pattern.

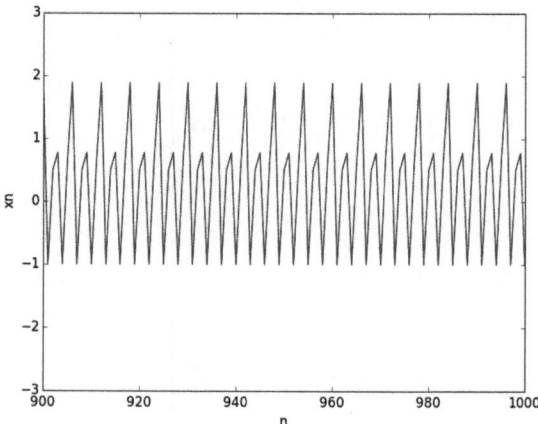

Fig. 6.3 An example of a periodic cycle with spiking several patterns as we can see spikes with different heights.

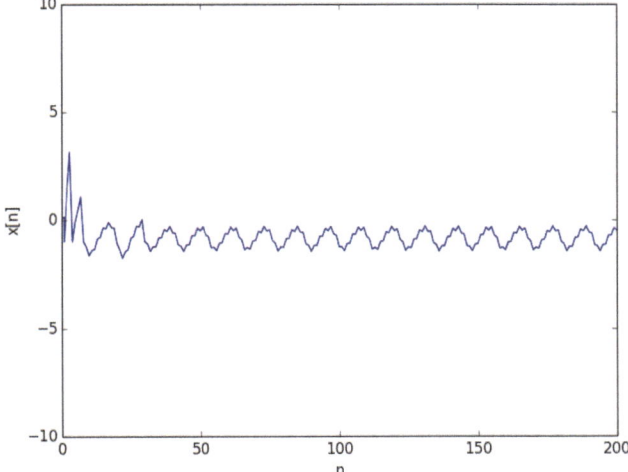

Fig. 6.4 An example of an eventually periodic cycle with several spiking patterns.

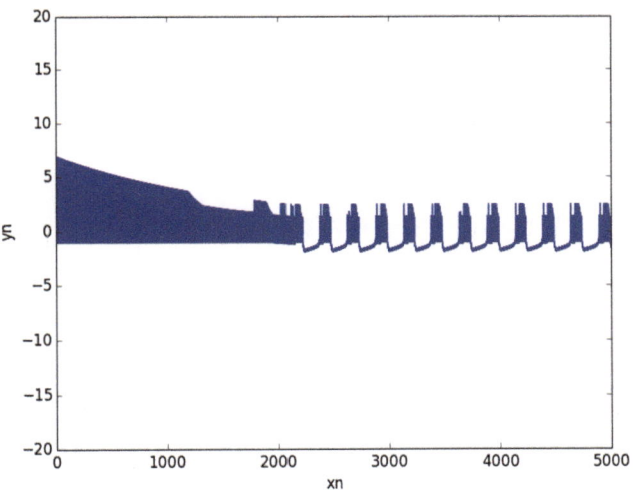

Fig. 6.5 An example of an eventually periodic cycle with several clusters of spiking patterns.

6.7 Additional Examples of Periodicity Graphs

We experienced various patterns of periodic cycles, eventually periodic solutions
and patterns of periodic cycles that are composed of clusters and of various patterns;
in particular, one pattern transforming to another pattern. For example, the graph
below (Figure 6.6):

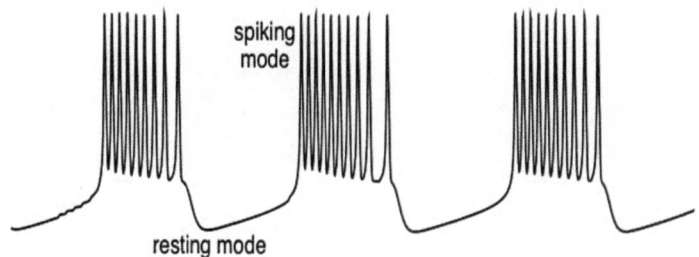

Fig. 6.6 An example of clusters of spikes separated from each other with a steady transition. This particular pattern was discovered in the **Izhikevich Neural Networking Model** [23]. More patterns of these spiking patterns can be found in [23].

We can also observe similar periodic behaviors in **Signal Processing in Sigma-Delta Domain** in [14] and [32]. Now we will switch gears by analyzing the graph below (Figure 6.7):

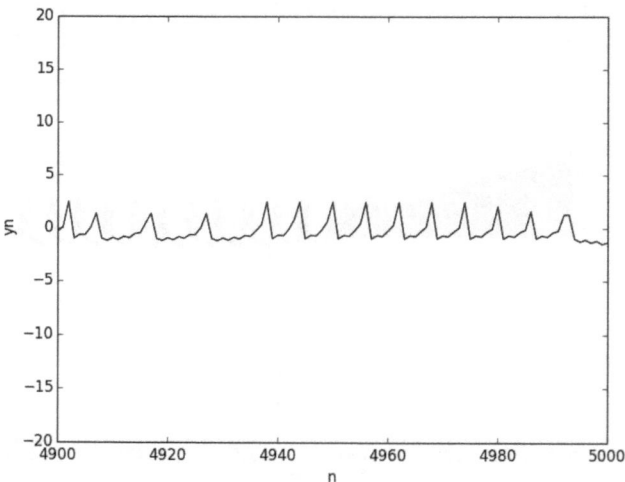

Fig. 6.7 An example of transition from one pattern to another pattern. However, the second pattern does not hold steady and then the pattern therefore becomes unstable. We have never noticed unstable patterns but they can certainly occur. This is where very carefully computer observations are vital to make such deductions.

We will conclude this chapter and the book with the following diagram (Figure 6.8):

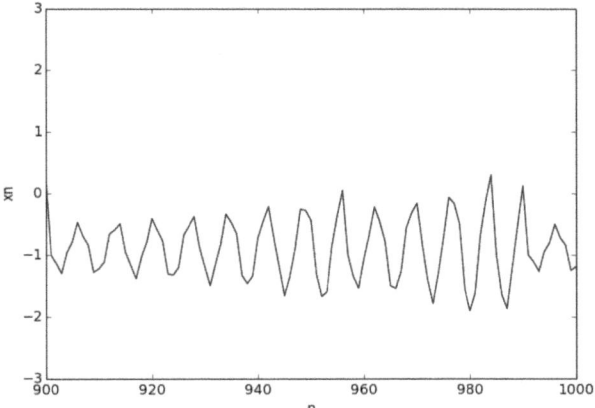

Fig. 6.8 An example of an unstable chaotic orbit.

These are examples of the **Rulkov Model** and can be found in [33] and [34].

Appendices

A.1 Patterns of Sequences

1. Linear Patterns:

$$1 , 2 , 3 , 4 , 5 , 6 , 7 , \ldots \ = \ \{n\}_{n=1}^{\infty}$$

$$2 , 4 , 6 , 8 , 10 , 12 , 14 , \ldots \ = \ \{2n\}_{n=1}^{\infty}$$

$$1 , 3 , 5 , 7 , 9 , 11 , 13 , \ldots \ = \ \{2n+1\}_{n=0}^{\infty}$$

$$3 , 6 , 9 , 12 , 15 , 18 , 21 , \ldots \ = \ \{3n\}_{n=1}^{\infty}$$

2. Quadratic Patterns:

$$1 , 4 , 9 , 16 , 25 , 36 , 49 , \ldots \ = \ \{n^2\}_{n=1}^{\infty}$$

$$4 , 16 , 36 , 64 , 100 , 144 , 196 , \ldots \ = \ \{(2n)^2\}_{n=1}^{\infty}$$

$$1 , 9 , 25 , 49 , 81 , 121 , 169 , \ldots \ = \ \{(2n-1)^2\}_{n=1}^{\infty}$$

3. Geometric Patterns:

© Springer Nature Switzerland AG 2018
M. A. Radin, *Periodic Character and Patterns of Recursive Sequences*,
https://doi.org/10.1007/978-3-030-01780-4

$$1, r, r^2, r^3, r^4, r^5, r^6, \ldots = \{r^n\}_{n=0}^{\infty}$$

$$2, 4, 8, 16, 32, 64, 128, \ldots = \{2^n\}_{n=1}^{\infty}$$

$$3, 9, 27, 81, 243, 729, , \ldots = \{3^n\}_{n=1}^{\infty}$$

A.2 Alternating Patterns of Sequences

1. Linear Alternating Patterns:

$$1, -2, 3, -4, 5, -6, 7, \ldots = \{(-1)^{n+1} n\}_{n=1}^{\infty}$$

$$-1, 2, -3, 4, -5, 6, -7, \ldots = \{(-1)^n n\}_{n=1}^{\infty}$$

$$1, -3, 5, -7, 9, -11, 13, \ldots = \{(-1)^n [2n+1]\}_{n=0}^{\infty}$$

$$-1, 3, -5, 7, -9, 11, -13, \ldots = \{(-1)^{n+1} [2n+1]\}_{n=0}^{\infty}$$

2. Quadratic Alternating Patterns:

$$1, -4, 9, -16, 25, -36, 49, \ldots = \{(-1)^{n+1} n^2\}_{n=1}^{\infty}$$

$$-1, 4, -9, 16, -25, 36, -49, \ldots = \{(-1)^n n^2\}_{n=1}^{\infty}$$

3. Geometric Alternating Patterns:

$$1, -r, r^2, -r^3, r^4, -r^5, r^6, \ldots = \{(-1)^n r^n\}_{n=0}^{\infty}$$

$$-1, r, -r^2, r^3, -r^4, r^5, -r^6, \ldots = \{(-1)^{n+1} r^n\}_{n=0}^{\infty}$$

A.3 Finite Series

$$1 + 2 + 3 + 4 + 5 + 6 + \ldots + n = \sum_{i=1}^{n} i = \frac{n[n+1]}{2} \, .$$

$$1 + 3 + 5 + 7 + 9 + 11 + \ldots + [2n-1] = \sum_{i=1}^{n} (2i-1) = n^2 \, .$$

$$1 + 4 + 9 + 16 + 26 + 36 + \ldots + n^2 = \sum_{i=1}^{n} i^2 = \frac{n[n+1][2n+1]}{6} \, .$$

$$1 \cdot 2 + 2 \cdot 3 + 3 \cdot 4 + 4 \cdot 5 + \ldots + n \cdot [n+1] = \sum_{i=1}^{n} i \cdot [i+1] = \frac{n[n+1][n+2]}{3} \, .$$

$$\frac{1}{1 \cdot 2} + \frac{1}{2 \cdot 3} + \frac{1}{3 \cdot 4} + \frac{1}{4 \cdot 5} + \ldots + \frac{1}{n \cdot [n+1]} = \sum_{i=1}^{n} \frac{1}{i \cdot [i+1]} = \frac{n}{n+1} \, .$$

$$1 + r + r^2 + r^3 + r^4 + r^5 + \ldots + r^n = \sum_{i=0}^{n} r^i = \frac{1-r^{n+1}}{1-r} \, .$$

$$1 \cdot 2^0 + 2 \cdot 2^1 + 3 \cdot 2^2 + 4 \cdot 2^3 + \ldots + n \cdot 2^{n-1} = \sum_{i=1}^{n} i \cdot 2^{i-1} = [n-1]2^n + 1 \, .$$

$$\binom{n}{0} + \binom{n}{1} + \binom{n}{2} + \ldots + \binom{n}{n-1} + \binom{n}{n} = \sum_{i=0}^{n} \binom{n}{i} = 2^n \, .$$

A.4 Convergent Infinite Series

$$1 + r + r^2 + r^3 + r^4 + r^5 + \ldots + = \sum_{n=0}^{\infty} r^n = \frac{1}{1-r} \, , \quad |r| < 1 \, .$$

$$\frac{1}{1 \cdot 2} + \frac{1}{2 \cdot 3} + \frac{1}{3 \cdot 4} + \frac{1}{4 \cdot 5} + \ldots + = \sum_{n=1}^{\infty} \frac{1}{n[n+1]} = 1 \, .$$

$$2 + \frac{1}{2} + \frac{1}{6} + \frac{1}{24} + \frac{1}{120} + \frac{1}{720} + \ldots + = \sum_{n=0}^{\infty} \frac{1}{n!} = e \, .$$

$$1 - \frac{1}{2} + \frac{1}{3} - \frac{1}{4} + \frac{1}{5} - \frac{1}{6} + \ldots + = \sum_{n=1}^{\infty} \frac{(-1)^{n+1}}{n} = Ln[2] \ .$$

$$1 + \frac{1}{4} + \frac{1}{9} + \frac{1}{16} + \frac{1}{25} + \frac{1}{36} + \ldots + = \sum_{n=1}^{\infty} \frac{1}{n^2} = \frac{\pi^2}{6} \ .$$

A.5 Periodic Sequences and Modulo Arithmetic

Period-2 sequence $\{A_n\}_{n=0}^{\infty}$ and the following period-2 pattern:

$$A_0 , A_1 , A_0 , A_1 , \ldots .$$

Period-2 sequence $\{A_n\}_{n=0}^{\infty}$ and the following period-2 pattern:

$$\frac{A_0 A_1 - 1}{1 + A_0} , \ \frac{A_0 A_1 - 1}{1 + A_1} , \ \frac{A_0 A_1 - 1}{1 + A_0} , \ \frac{A_0 A_1 - 1}{1 + A_1} , \ldots .$$

Period-3 sequence $\{A_n\}_{n=0}^{\infty}$ and the following period-3 pattern:

$$A_0 , A_1 , A_2 , A_0 , A_1 , A_2 , \ldots .$$

Period-3 sequence $\{A_n\}_{n=0}^{\infty}$ and the following period-3 pattern:

$$\frac{A_0 A_1}{A_0 A_1 A_2 + 1} , \ \frac{A_1 A_2}{A_0 A_1 A_2 + 1} , \ \frac{A_2 A_0}{A_0 A_1 A_2 + 1} , \ldots .$$

Period-4 sequence $\{A_n\}_{n=0}^{\infty}$ and the following period-4 pattern:

$$A_0 , A_1 , A_2 , A_3 , A_0 , A_1 , A_2 , A_3 , \ldots .$$

Period-4 sequence $\{A_n\}_{n=0}^{\infty}$ and the following period-4 pattern:

$$\frac{A_0 + A_1 + A_2}{2} , \frac{A_1 + A_2 + A_3}{2} , \frac{A_2 + A_3 + A_0}{2} , \frac{A_3 + A_0 + A_1}{2} , \cdots .$$

A.6 Alternating Periodic Sequences and Modulo Arithmetic

Period-2 sequence $\{A_n\}_{n=0}^{\infty}$ and the following period-4 pattern:

$$A_0 , A_1 , -A_0 , -A_1 , \cdots .$$

Period-3 sequence $\{A_n\}_{n=0}^{\infty}$ and the following period-6 pattern:

$$A_0 , A_1 , A_2 , -A_0 , -A_1 , -A_2 , \cdots .$$

Period-2 sequence $\{A_n\}_{n=0}^{\infty}$ and the following period-2 pattern:

$$\frac{A_0 A_1}{A_0 + A_1 + 1} , \frac{-A_0 A_1}{A_0 + A_1 + 1} , \cdots .$$

Period-2 sequence $\{A_n\}_{n=0}^{\infty}$ and the following period-2 pattern:

$$\frac{A_0 - A_1}{A_0 A_1 + 1} , \frac{A_1 - A_0}{A_0 A_1 + 1} , \cdots .$$

Period-2 sequence $\{A_n\}_{n=0}^{\infty}$ and the following period-4 pattern:

$$\frac{A_0}{A_0 A_1 + 1} , \frac{A_1}{A_0 A_1 + 1} , \frac{-A_0}{A_0 A_1 + 1} , \frac{-A_1}{A_0 A_1 + 1} , \cdots .$$

Period-2 sequence $\{A_n\}_{n=0}^{\infty}$ and the following period-4 pattern:

$$\frac{A_0 + A_1}{A_0 A_1 + 1} , \frac{A_0 - A_1}{A_0 A_1 + 1} , \frac{-[A_0 + A_1]}{A_0 A_1 + 1} , \frac{A_1 - A_0}{A_0 A_1 + 1} , \cdots .$$

References

1. A.M. Amleh, D.A. Georgiou, E.A. Grove, and G. Ladas, On the recursive sequence $x_{n+1} = \alpha + \frac{x_{n-1}}{x_n}$, *J. Math. Anal. Appl.*; (233)(1999), 790–798.

2. A.M. Amleh, J. Hoag, and G. Ladas, A Difference Equation with Eventually Periodic Solutions, *Computer Math. Applic. 36*, (1998), 401–404.

3. Anisimova, A., Avotina, M., Bula, I., Periodic Orbits of Single Neuron Models with Internal Decay Rate $0 < \beta \leq 1$. *Mathematical Modelling and Analysis* **18**, (2013), 325–345.

4. Anisimova, A., Avotina, M., Bula, I., Periodic and Chaotic Orbits of a Neuron Model. *Mathematical Modelling and Analysis* **20**, 30–52 (2015)

5. W.F. Basener, B.P. Brooks, M.A. Radin, and T. Wiandt, Rat Instigated Human Population Collapse on Easter Island. *Journal of Non-Linear Dynamics, Psychology and Life Sciences 12*, **3**, 227–240 (2008).

6. W.F. Basener, B.P. Brooks, M.A. Radin, and T. Wiandt, Dynamics of a Population Model for Extinction and Sustainability in Ancient Civilizations. *Journal of Non-Linear Dynamics, Psychology and Life Sciences 12*, **1**, 29–54 (2008).

7. W.J. Briden, E.A. Grove, C.M. Kent, and G. Ladas, Eventually Periodic Solutions of $x_{n+1} = max\{\frac{1}{x_n}, \frac{A_n}{x_{n-1}}\}$, *Commun. Appl. Nonlinear Anal. 6* (1999), no. 4.

8. W.J. Briden, E.A. Grove, G. Ladas, and L.C. McGrath, On the Nonautonomous Equation $x_{n+1} = max\{\frac{A_n}{x_n}, \frac{B_n}{x_{n-1}}\}$, *Proceedings of the Third International Conference on Difference Equations and Applications*. September 1–5, 1997, Taipei, Taiwan, Gordon and Breach Science Publishers (1999), 49–73.

9. W.J. Briden, G. Ladas, and T. Nesemann, On the Recursive Sequence $x_{n+1} = max\left\{\frac{1}{x_n}, \frac{A_n}{x_{n-1}}\right\}$, *J. Differ. Equations. Appl. 5* (1999), 491–494.

© Springer Nature Switzerland AG 2018
M. A. Radin, *Periodic Character and Patterns of Recursive Sequences*,
https://doi.org/10.1007/978-3-030-01780-4

10. Bula, I., Radin, M.A., Wilkins, N., Neuron model with a period three internal decay rate. *Electronic Journal of Qualitative Theory of Differential Equations* (EJQTDE); (46), 1–19 (2017).

11. Bula, I., Radin, M.A., Periodic orbits of a neuron model with periodic internal decay rate. *Applied Mathematics and Computation* (266), 293–303 (2015).

12. E. Camouzis, G. Ladas, I.W. Rodriques, and S. Northsfield. On the rational recursive sequences $x_{n+1} = \frac{\beta x_n^2}{1+x_n^2}$. *Computers Math. Appl.*, 28:37–43, 1994.

13. E. Chatterjee, E.A. Grove, Y. Kostrov, and G. Ladas, On the Trichotomy character of $x_{n+1} = \frac{\alpha+\gamma x_{n-1}}{A+Bx_n+x_{n-2}}$, *J. Difference Equa. Appl.*,(9)(2003), 1113–1128.

14. V. Da Fonte Dias, Signal Processing in the Sigma - Delta Domain, *Microelectronics Journal*, (26) (1995), 543–562.

15. J. Diblik, M. Ruzickova, and E.L. Schmeidel, Existence of asymptotically periodic solutions of system of Volterra difference equations, *Journal of Difference Equations and Applications*, 15 (11–12)(2009), 1165–1177.

16. J. Diblik, M. Ruzickova, and E.L. Schmeidel, Asymptotically periodic solutions of system of Volterra difference equations, *Computers and Mathematics with Applications*, 59(2010), 2854–2867.

17. J. Diblik, M. Ruzickova, E.L. Schmeidel and M. Zbaszyniak, Weighted Asymptotically Periodic Solutions of Linear Volterra Difference Equations, *Hindawi Publishing Corporation Abstract and Applied Analysis*, (2011).

18. K.R. Janglajew, E.L. Schmeidel, Periodicity of solutions of nonhomogeneous linear difference equations, *Advances in Difference Equations* (2012), 195.

19. C. Gibbons, M.R.S. Kulenovic, and G. Ladas, *On the Recursive Sequence*

$$x_{n+1} = \frac{\alpha + \beta x_{n-1}}{\gamma + x_n} ,$$

Math. Sci. Res. Hot-Line 4 (2) (2000), 1–11.

20. E.A. Grove, G. Ladas, M. Predescu, and M. Radin, On the global character of $x_{n+1} = \frac{px_n+n_{n-1}}{q+x_n}$,*Mathematical Science Research Hot-Line*,5(7)(2001), 25–39.

21. E.A. Grove, G. Ladas, M. Predescu, and M. Radin, On the global character of the difference equation $x_{n+1} = \frac{\alpha+\gamma x_{n-(2k+1)}+\delta x_{n-2l}}{A+x_{n-2l}}$, *J. Difference Equa. Appl.*, (9)(2003), 171–200.

22. E.A. Grove, E. Lapierre, and W. Tikjha, On the global behavior of $x_{n+1} = |x_n|-y_n-1$ and $y_{n+1}=x_n+|y_n|$, *Cubo A Mathematica Journal*, (14)(2)(2012), 111–152.

23. E.M. Izhikevich, Simple Model of Spiking Neurons, *IEEE Transactions on Neural Networks*, (14)(6)(2003), 1569–1572.

24. G.L. Karakostas and S. Stevic, On the recursive sequence $x_{n+1} = B + \frac{x_{n-k}}{a_0 x_n + \cdots + a_{k-1} x_{n-k+1} + \gamma}$, *J. Difference Equa. Appl.*, (10)(2004), 809–815.

25. C.M. Kent, E.A. Grove, G. Ladas, and M.A. Radin; On $x_{n+1} = max \left\{ \frac{1}{x_n}, \frac{A_n}{x_{n-1}} \right\}$ with a Period 3 Parameter; *Fields Institute Communications, Volume 29, March 2001*.

26. C.M. Kent, M. Kustesky, A.Q. Nguyen, B.V. Nguyen, Eventually Periodic Solutions of $x_{n+1} = max \left\{ \frac{A_n}{x_n}, \frac{B_n}{x_{n-1}} \right\}$, With Period Two Cycle Parameters.

27. M.R.S. Kulenovic, G. Ladas, and N.R. Prokup, On the Recursive Sequence

$$x_{n+1} = \frac{a x_n + b x_{n-1}}{A + x_n}.$$

J. Differ. Equations Appl. 6(5) (2000), 563–576.

28. M.R.S. Kulenovic, G. Ladas, and W.S. Sizer, *On the Recursive Sequence*

$$x_{n+1} = \frac{\alpha x_n + \beta x_{n-1}}{\gamma x_n + \delta x_{n-1}}$$

Math. Sci. Res. Hot-Line 2 (5) (1998), 1–16.

29. C.M. Kent, M.A. Radin; On the Boundedness of Positive Solutions of a Reciprocal Max-Type Difference Equation with Periodic Parameters; *International Journal of Difference Equations*, (8)(2)(2013), 195–213.

30. G. Ladas, On the recursive sequence $x_{n+1} = max \left\{ \frac{A_0}{x_n}, \ldots, \frac{A_k}{x_{n-k}} \right\}$, *J. Diff. Equa. Appl.*, **2** (2) (1996), 339–341.

31. E. Lapierre, Y. Lenbury, and W. Tikjha, On the global character of the system of Piecewise Linear Difference Equations $x_{n+1} = |x_n| - y_n - 1$ and $y_{n+1} = x_n + |y_n|$, *Hindawi Publishing Corporation, Advances in Difference Equations*, (2010).

32. M.A. Pervez, H.V. Sorensen, and J. Van der Spiegel, An overview of Sigma-Delta Converters, *IECE Signal Processing Magazine*, (1996), 61–84.

33. Pisarchik, A.N., Radin, M.A., Vogt, R., Nonautonomous Discrete Neuron Model with Multiple Periodic and Eventually Periodic Solutions. *Discrete Dynamics in Nature and Society* **2015**, Article ID 147282, 6 pages (2015)

34. N.F. Rulkov, Modeling of spiking-bursting neural behavior using two-dimensional map. *Physical Review E*, V. 65, P. 041922, (2002).

35. S. Stević, On the recursive sequence $x_{n+1} = \frac{\alpha + \sum_{i=1}^{k} \alpha_i x_{n-p_i}}{1 + \sum_{j=1}^{m} \beta_j x_{n-q_j}}$, *J. Difference Equa. Appl.*, (13)(2007), 41–46.

36. Zhou, Z., Wu, J., Stable Periodic Orbits in Nonlinear Discrete-Time Neural Networks with Delayed Feedback. *Computers and Mathematics with Applications* **45**, 935–942 (2003)

37. Zhou, Z., Periodic Orbits on Discrete Dynamical Systems. *Computers and Mathematics with Applications* **45**, 1155–1161 (2003)

38. Zhu, H., Huang, L., Dynamics of a Class of Nonlinear Discrete-Time neural Networks. *Computers and Mathematics with Applications* **48**, 85–94 (2004)

39. Yuan, Z., Huang, L., Chen, Y., Convergence and Periodicity of Solutions for a Discrete-Time Network Model of Two Neurons. *Mathematical and Computer Modelling* **35**, 941–950 (2002)

40. Yuan, Z., Huang, L., All solutions of a class of discrete-time systems are eventually periodic. *Applied Mathematics and Computation* **158**, 537–546 (2004)

Index